Philosophy of Engineering and Technology

Volume 44

The *Philosophy of Engineering and Technology* book series provides the multifaceted and rapidly growing discipline of the philosophy of technology with a central overarching and integrative platform. It publishes on all topics in the philosophy of technology and is open to all research communities across the world.

Specifically, it publishes edited volumes and monographs in:

- the phenomenology, anthropology and socio-politics of technology and engineering
- the emergent fields of the ontology and epistemology of technology and design
- engineering ethics and the ethics of specific technologies ranging from nuclear technologies to artificial intelligence
- written from philosophical and practitioners' perspectives and authored by philosophers and practitioners

The series also welcomes proposals that bring these fields together or advance philosophy of engineering and technology in other integrative ways.

Proposals should include:

- A short synopsis of the work or the introduction chapter
- The proposed Table of Contents

- The CV of the lead author(s) or editor(s)
- If available: one sample chapter

We aim to make a first decision within 1 month of submission. In case of a positive first decision the work will be provisionally contracted; the final decision about publication will depend upon the result of anonymous peer review of the complete manuscript. We aim to have the complete work peer-reviewed within 3 months of submission.

The series discourages the submission of manuscripts that contain reprints of previous published material and/or manuscripts that are below 150 pages/75,000 words.

For inquiries and submission of proposals authors can contact the editor-in-chief Pieter Vermaas via: p.e.vermaas@tudelft.nl, or contact one of the associate editors.

Merel Noorman • Mario Verdicchio
Editors

Computer Ethics Across Disciplines

Deborah G. Johnson and Algorithmic Accountability

Springer

Editors
Merel Noorman
Tilburg University
Tilburg, Limburg, The Netherlands

Mario Verdicchio
University of Bergamo
Bergamo, Italy

ISSN 1879-7202 ISSN 1879-7210 (electronic)
Philosophy of Engineering and Technology
ISBN 978-3-031-87973-9 ISBN 978-3-031-87974-6 (eBook)
https://doi.org/10.1007/978-3-031-87974-6

This Springer imprint is published by the registered company Springer Nature Switzerland AG
The registered company address is: Gewerbestrasse 11, 6330 Cham, Switzerland

Contents

Chapter 1
Editors' Introduction

Merel Noorman and Mario Verdicchio

Abstract Computer Ethics is a practical kind of philosophy that is aimed at investigating how computer technologies should be used. It has accompanied the development of these technologies for over half a century. One of the leading figures in this field has been Deborah G. Johnson, whose agenda-setting handbook on Computer Ethics has inspired generations of scholars to explore the new ethical questions that these technologies raise. In 2021, Johnson was awarded the Society of Philosophy of Technology Lifetime Achievement Award for her outstanding contribution to computer and engineering ethics. In recognition of this award and her research on Computer Ethics, this edited volume brings together philosophers and scholars from other disciplines including computer scientists, cognitive scientists, and STS scholars, who have engaged with Johnson's extensive body of work. Some of the contributors, such as van den Hoven and Miller, have helped shape the field of computer ethics, while contributors from later generations who benefitted from the work of these trailblazers are following their path and further exploring and expanding on their legacy. The volume seeks to introduce the lessons learned from Computer Ethics to a broader audience of scholars from different disciplines and show how they still resonate in today's ethical discussions about new emerging computer technologies. Each chapter illustrates how combining philosophy of technology, ethics, and different disciplinary perspectives can help analyze and clarify the complex intricacies of computer technology and societies. The common thread in these chapters is a focus on issues of algorithmic accountability. As the only (for

M. Noorman (✉)
Tilburg Institute for Law Technology and Society, Tilburg University,
Tilburg, The Netherlands
e-mail: m.e.noorman@tilburguniversity.edu

M. Verdicchio
Department of Foreign Languages, Literatures, and Cultures,
University of Bergamo, Bergamo, Italy

Berlin Ethics Lab, Technische Universität Berlin, Berlin, Germany
e-mail: mario.verdicchio@unibg.it

M. Noorman, M. Verdicchio (eds.), *Computer Ethics Across Disciplines*,
Philosophy of Engineering and Technology 44,
https://doi.org/10.1007/978-3-031-87974-6_1

now) full-fledged moral agents, humans are called to action to discuss, propose, negotiate, and implement ethical ways to use computers, which obviously and most importantly include attribution of responsibility when something goes wrong. This introductory chapter situates Johnson's work within the broader discussion on Computer Ethics and provides short summaries of the contributions to this volume.

Keywords Computer ethics · Algorithmic accountability · Deborah G. Johnson

Computer ethics, as a field of research, emerged in the second half of the last century alongside the rapid development of computer technologies. As computer technologies became part of everyday activities, questions emerged about the ethical use of these technologies that provided new capabilities and new choices for action (Moor, 1985). Questions about, for example, to what extent employers could use computers to monitor the activities of their employees or whether users could be held responsible for the spread of computer viruses. Since those early days, Computer Ethics has continued to develop as a field of research concerned with a wide variety of ethical challenges that new computer technologies bring.

One of the early pioneers of this field is Professor Deborah G. Johnson. In her research, she has extensively explored the ethical implications of increasingly autonomous and opaque computer technologies and the responsibilities of technology professionals. Her research is positioned on the intersection of technology, policy making, and ethics, examining how advancements in technology influence society and affect individual rights. In 2021, Johnson was awarded the Society of Philosophy of Technology Lifetime Achievement Award for her outstanding contribution to computer and engineering ethics. In recognition of this award and her research on Computer Ethics, this edited volume brings together philosophers and scholars from other disciplines including computer scientists, cognitive scientists, and STS scholars, who have engaged with Johnson's extensive body of work. The central theme of the book is inspired by Johnson's keynote address on algorithmic accountability that she presented at the occasion of the award. In this introductory chapter, we first provide some historical background on the field of Computer Ethics and the role that Johnson has played in shaping the field. We will then give an outline of the rest of the volume.

Placing Human Beings at the Center

Algorithmic accountability, as a term, falls into a category of adjective-noun binomials where an endeavor, a phenomenon, or, more in general, a facet of human culture gets characterized by a technological and, often, computational dimension. Many other examples come to mind, like Computer Art, Artificial Intelligence, Machine Learning, and, more significantly in this context, Computer Ethics. Indeed, algorithmic accountability, referring to the responsibility in the conception, development, deployment, and consequent impact of algorithms used in decision-making processes, can be considered a specific part of the more general field of Computer

Ethics, a practical kind of philosophy that is aimed at investigating how computer technology should be used.

Computer Ethics has often been defined as ethics applied to the computer world (Brey, 2000; Floridi, 1999; Moor, 1985). However, this is a far too simple way to describe the field. In fact, combining an adjective with a noun does not always specify well how the relevant conceptual characterization occurs, especially when the words in the pairing have complex meanings that can each prompt discussion. Simpler combinations like "red flag" do not pose any problems, whereas a term like "Computer Ethics" is much more challenging because of all the questions that arise at the intersection between the two vast contexts involved. These questions are ontologic, epistemic, disciplinary, sociological, and technological. Ontology guides us in our explorations of what exists. We wonder what entities we are dealing with, whether they can be categorized, and how, and what relationships hold between them. This is the perspective that lays the ground for all further endeavors because it is an attempt at establishing the fundamental concepts and definitions that we are going to use to develop a discourse encompassing computers on the one side, and ethics on the other. The crucial question here revolves around the nature of these entities, which we investigate to understand how we are to combine them to make, specialize, or transform Ethics into Computer Ethics.

From these ontological foundations, whether solid or shaky, all the other questions derive, epistemic questions about the kind of knowledge that we can acquire in this effort; disciplinary questions about the boundaries and the intersections of the disciplines we are building with such knowledge; sociological questions about how the relevant concepts, discourses, and prescriptions are elaborated and processed within a social framework, and, finally, technological questions on how these intellectual activities shape and, in turn, are shaped by a parallel yet entangled development of technological artifacts.

It soon becomes evident that an endeavor punctuated by all these investigations cannot be aptly described by a mere adjective-noun binomial. Indeed, another instance, namely Machine Ethics, is comprised of almost the same terms as Computer Ethics and yet it indicates a profoundly different approach, based on a markedly distinct ontology. Machine Ethics proposes a realm where ethical considerations are embedded within the very fabric of computational artifacts (Anderson & Anderson, 2011; Nath & Sahu, 2020). Taking into account how such machines work in terms of software and hardware, this means that Machine Ethics navigates the terrain of moral agency and decision-making algorithms with the debatable assumption that all things ethical, i.e. principles, criteria, implications, judgments, etc., are susceptible to a process of computational modeling that makes them compatible with a computer's inner workings.

Deborah Johnson has been debating this supposition strenuously for decades, starting with the publication of "Computer Ethics" in 1985, the very first textbook on the topic and the reason why Johnson is considered one of the pioneers of this discipline. Back then, there was no explicit debate yet with supporters of Machine Ethics: the term would be coined only two years later. Johnson's mark, however, was already very clear: the perspective of Computer Ethics was centered on human beings. The discipline was about how humans are to develop and use computers in an ethical way, and not about how to build computers to make them act ethically: throughout Johnson's work, full-fledged agency has always been held tightly in human hands. Even when involved in an ethically charged context where humans and machines interact and where the latter are programmed to exhibit very articulated and nuanced behaviors, artifacts can only be moral entities, but not moral agents (Johnson, 2006). This was indeed one of the primary takeaways of Johnson's textbook, which also helped consolidate the framework of the discipline: computers entering many domains of human life were not to be seen as bringing in a new ethics, but a new field for the application of the already existing one. Humans have been tackling ethical issues for centuries if not millennia, and so were they to continue, with a "new twist" of having to include issues related to computational artifacts.

This position is what has led several scholars to frame Johnson as an "instrumentalist", that is, as somebody who embraces the view according to which computers, as sophisticated as they may be, are mere tools or instruments of human action (Gunkel, 2020). This is not inherently wrong: Johnson does draw an ontological line between humans and machines - especially in her more recent work on Artificial Intelligence and Machine Learning - as a warning against the excessively liberal use of some concepts, like agency or autonomy. These disciplines, interestingly, carry similar issues in their very names (Johnson & Noorman, 2013; Johnson & Verdicchio, 2023). However, an "instrumentalist" tag is oversimplifying and reductive of Johnson's fundamental contribution to the discourse on ethics and technology because it only highlights what is conceptually taken away from artifacts or, perhaps, what is argued to be absent in artifacts. It does not show what is attributed to them, by whom such attributions are made, and with what consequences, in a truly authentic sociotechnical fashion.

Johnson has always been a champion of the sociotechnical paradigm. This paradigm prescribes that both the social and technical aspects of a complex system be taken into account as interdependent (Bijker & Law, 1994). It sheds light on the shortcomings of an "instrumentalist" interpretation of her work. Not recognizing agency or autonomy as part of technological artifacts in the way we do with humans does not reduce them to mere tools. We are well aware, also thanks to Johnson's help, of the complex network of decisions and agreements among people that is necessary to create, deploy, and maintain those artifacts and, in turn, of the very significant influence that artifacts exert on people in terms of changes in conceptualizations, relations, and values. Even without agency, technological artifacts are not mere tools, because of their role in affecting how humans make sense of themselves, of the artifacts they create, and of all the relations between them. From this perspective, Computer Ethics and Machine Ethics can themselves be conceptualized as a disciplinary embodiment of the sociotechnical impact of computers on our collective thinking.

Still, this narrative is not meant to draw a perfectly symmetrical picture including humans and computers: even if they all contribute in a fundamental way to the discourse on ethics. If only humans can be moral agents, then the focus is clearly on them when it comes to decisions, consequences, and responsibilities. This view, if interpreted in an atemporal, purely ontological sense, seems to pit Computer Ethics and Machine Ethics against each other on the issues of human and artificial moral agency. From a more contingent and contemporary perspective, debates on computers as moral agents may be postponed while awaiting groundbreaking technological developments that may corroborate the stances of computationalism that still appear far-fetched today. A human-centric approach draws attention to what human beings do with existing artifacts and what ethical questions are raised in practice: a call that aligns with the emphasis of Computer Ethics on a pragmatic approach to philosophy.

Computer Ethics as a Practice-Oriented Field

Since its early beginnings, Computer Ethics has focused on concrete cases to understand the ethical questions that new computer technologies raised. Even before the term was coined, Nobert Wiener - as one of the early scholars to call attention to the ethics of computer technology - highlighted concrete cases regarding the integration of computer technology into society that affected life, health, security, happiness, freedom, knowledge, opportunities, or other key human values (Bynum, 2008). Later scholars such as Walter Maner, James Moor, and Johnson would do the same (Johnson, 1994; Moor, 1985; Walter, 1980). The methodology that they embraced was to look for the issues in these cases, clarify any ambiguous concepts or principles that seemed to be at stake and explore to what extent existing ethical principles, tools, concepts, or strategies would apply. This methodology, in various forms and shapes, has remained central to Computer Ethics. Since this early work, scholars have examined questions that computer technologies raised concerning privacy, equality, participation, email spam, technological dependence, cyberstalking, professional responsibility, machine autonomy, and much more (Adam, 2001; Borenstein et al., 2017; Gotterbarn, 1991; Himma & Tavani, 2008; Nissenbaum, 2004).

Within the field, opinions differed about whether the issues accompanying emerging computer technologies were unique and required a new kind of ethics, for example in addition to or in analogy with medical ethics or business ethics. Was there something fundamentally different about these technologies and their interaction with human beings and society that could not be addressed by received ethical theories, such as consequentialism or deontology? Did they require new ethical concepts or theories? Since the mid-twentieth century computers have undergone several extraordinary transformations, in terms of connectivity, performance, scale, and fields of application that presented new questions, such as are social media companies responsible for harmful content on their platforms or should machines be held accountable? Some scholars have suggested that traditional ethical theories were not equipped to deal with such questions and that new foundations were needed (Floridi, 1999; Maner, 1996). The ongoing developments in the field of computing also led some to suggest that computer ethics was surpassed by these developments and that other adjective-noun binomials were required, such as information ethics, internet ethics, cyberethics, and most recently AI ethics (Gordon & Nyholm, 2021). Much of what is currently discussed about the ethics of AI, such as the need for ethical guidelines and the disruptive effects of deep learning, is familiar territory for computer ethicists. It echoes earlier discussions on changing social practices and norms, interpretations of moral concepts and principles, the need for professional responsibility and codes of ethics, etc. The emerging fields of ethics highlight different aspects of digital, computational, and information technologies and have served different purposes, but they all share common roots in Computer Ethics.

Regardless of the name, Computer ethics has provided a methodology, concepts, tools, and strategies to deal with new developments in the field of computing that are still relevant today. In her handbook, Johnson argued that computers "pose new versions of standard moral problems and moral dilemmas, exacerbating the old problems, and forcing us to apply ordinary moral norms in uncharted realms" (Johnson, 1994). This position directs the attention to interactions between the development and introduction of new computer technologies. It requires us to understand how technologies and ethics interact, and what their components and moving parts are. It raises questions about how these technologies change the way human beings do things and how they relate to each other; how tasks and responsibilities are redistributed with the introduction of these technologies; or how they affect how we weigh and interpret moral values and translate them into norms. Since the publication of Johnson's agenda-setting textbook, a significant part of the field of computer ethics continues to take a practice-oriented approach to systematically sorting out how computer technologies interact with human beings, society, and the world and how that affects moral values and norms. Through case analyses, computer ethicists interrogated the various elements of the relationships within sociotechnical systems, including concepts, activities, and affordances. Computer ethicists clarify conflicts and choices and highlight the blind spots and limitations of certain perspectives or theoretical approaches. They apply existing ethical theories and concepts to new situations and assess whether they still suffice or if they need refinement, extension, or perhaps replacement.

To unpack ethical issues accompanying computer technologies, computer ethicists have drawn on other disciplines to better understand the relationship between human beings, technologies, and society. The connection with computer science is obvious. However, sociological insights have also contributed to understanding the changes in values and norms that come with developments in computer technologies. And, as some of the contributions in this volume show, other fields such as human factors research and organization studies have also provided tools, methods, models, and frameworks to better understand what constitutes particular ethical concerns. Johnson has been a leading figure in building on the insights from another relatively new field: science and technology studies (STS). This field of study provided insights into the mutual shaping of technology and society that helped to understand the interplay between technological change and morality (Brey, 2000; Van den Hoven et al., 2012). With her sharp focus on the social and political issues deeply entangled with computer technology, Johnson was at the forefront of computer ethicists shedding light on the many ways in which technology and society affect each other.

Overview of the Book

The chapters in this volume show how the lessons learned from Computer Ethics still resonate in today's ethical discussions about new emerging computer technologies. Some of the contributors such as van den Hoven and Miller have helped shape the field, while contributors from later generations who benefitted from the work of these trailblazers are following their path and further exploring and expanding on their legacy. The common thread in these chapters is a focus on issues of algorithmic accountability. As the only (for now) full-fledged moral agents, humans are called to action to discuss, propose, negotiate, and implement ethical ways to use computers, which obviously, and most importantly, include attribution of responsibility when something goes wrong. Moreover, each chapter illustrates how combining philosophy of technology, ethics, and different disciplinary perspectives can help analyze and clarify the complex intricacies of computer technology and societies.

In line with the Computer Ethics tradition, all the contributions to this volume are characterized by a strong connection to the real world and contain concrete, actionable proposals on how to do ethics in relation to computers. Furthermore, they all draw attention to interconnections between the different elements of sociotechnical systems. Johnson herself opens the book staying true to her stance: as she then defined Computer Ethics as a new twist on traditional ethics, so she argues now that to understand algorithmic accountability we have to go back to established accounts of accountability and examine how they can be extended to algorithmic decision-making. This chapter is an extended version of her keynote address for the Society of Philosophy and Technology on the occasion of receiving the award. She draws on the work of Mark Bovens to develop an account of accountability as a social practice consisting of actors and forums. She argues that a better understanding of how accountability is constituted as social practice can provide insights into how accountability gaps emerge in algorithmic decision-making as a result of the choices that human actors make.

In the following chapter, Keith Miller illustrates the significance of Johnson's work both in the field of computer ethics and in computer science as he provides an overview of Johnson's work on algorithmic accountability and explores what computer scientists can learn from it. More specifically, he outlines the different socio-technical aspects of algorithmic accountability as Johnson describes them and subsequently maps the concepts of actors, forums, and sanctions onto the world of computing. Through his analysis, he highlights her "passion about justice and human thriving". In Chap. 4, Kirsten Martin builds on Johnson's understanding of accountability as social practice and applies it to different setting to explore how accountability gaps can be addressed. She relates the notion of accountability practices to business ethics to shed light on why, to whom, and for what AI firms, in particular platform companies, are accountable. With this analysis, she critiques claims that these firms are not accountable for the moral implications of their decisions. Like Johnson, she shows that through a closer look at what constitutes accountability the various moving parts of accountability relationships can be brought into view to clarify the choices that can and are made in the redistribution of responsibility around new technologies.

The following two chapters pick up the theme of negotiations around moral values and norms that Johnson highlighted in her chapter. They zoom in on the normative dimensions of the embedding of computer technologies in society and examine how values and norms are constituted and contested within the context of AI. Jonne Maas and Jeroen van den Hoven give an illustration of some of the insights that computer ethics, STS, and political theory can provide for the emerging field of AI Ethics in Chap. 5. In particular, they show that the narrow focus in AI Ethics on aligning AI technologies with supposedly commonly accepted values and principles overlooks the societal, economic, and political complexity of the context in which these technologies are developed and used. They unpack the multiple layers of the political naivete exhibited by parts of the AI Ethics field, to argue for a broader focus on how "the system has come to be part of our social world." More specifically, they contend that AI Ethics should include a broader perspective on political-economic background that informs AI development, deployment, and ethical assessment. In Chap. 6, Merel Noorman looks at the broader governance context in which moral values and norms are negotiated as part of the transformations that new computer technologies bring. The emerging field of AI Governance provides a case in which such negotiations can be examined in more detail. Building on the practice theory as developed in STS, she shows that a practice-oriented approach to empirical philosophy can help understand how and when moral values and norms become subject to debate and how such debates are part of negotiations about what constitutes 'effective governance practices.' This offers opportunities, she argues, for empirical philosophers and ethicists to not only analyze such practices but also to intervene in these practices.

Computer Ethics has traditionally interrogated the different sociotechnical aspects of ethical issues through drawing on other disciplines. In Chap. 7, Mario Verdicchio turns the focus to the role of algorithms in the broader sociotechnical context in which accountability issues arise. He brings the discourse back to the fundamentals of computing and tries to map real cases of accountability to the workings of algorithms, programs, and computers. Through carefully tracing the relationships between these different elements of complex sociotechnical networks, he shows, how the term 'algorithmic accountability' may be a misnomer as algorithms are just one part of a much bigger whole. Nicholas Diakopoulos directs the attention in Chap. 8 the other way, towards the social side of accountability in sociotechnical systems, to highlight the normative work that needs to be done upfront to distribute responsibilities and establish expectations between actors in relation to algorithmic systems. He points to the underexplored role of news media in giving shape to and sustaining this kind of forward-looking accountability for algorithms.

In the final chapter, Thomas Powers provides a more critical perspective as he explores the limits of the sociotechnical approach taken by organization studies, engineering, and philosophy. He argues that such approaches are not able to provide the standard of clarity that explanations of technological failures require. Moreover, they have limited predictive power. He then turns the tables and asks if and how new computational technologies can help to better understand sociotechnical systems. In making this move, he also offers a perspective from which even within Johnson's understanding of accountability practices there is a possibility to hold computers accountable. In pointing at the necessity to make sociotechnical systems more understandable to a wider audience, Powers circles back to what Johnson has been arguing for: it is up to us, human beings.

References

Adam, A. (2001). Computer ethics in a different voice. *Information and Organization, 11*(4), 235–261.

Anderson, M., & Anderson, S. L. (2011). *Machine ethics*. Cambridge University Press.

Bijker, W. E., & Law, J. (1994). *Shaping technology/building society: Studies in sociotechnical change*. MIT Press.

Borenstein, J., Herkert, J., & Miller, K. (2017). Self-driving cars: Ethical responsibilities of design engineers. *IEEE Technology and Society Magazine, 36*(2), 67–75.

Brey, P. (2000). Disclosive computer ethics. *ACM Sigcas Computers and Society, 30*(4), 10–16.

Bynum, T. W. (2008). Milestones in the history of information and computer ethics. In *The handbook of information and computer ethics* (pp. 25–48).

Floridi, L. (1999). Information ethics: On the philosophical foundation of computer ethics. *Ethics and Information Technology, 1*(1), 33–52.

Gordon, J. S., & Nyholm, S. (2021). Ethics of artificial intelligence. *Internet Encyclopedia of Philosophy*. https://iep.utm.edu/ethics-of-artificial-intelligence/

Gotterbarn, D. (1991). Computer ethics: Responsibility regained. *National Forum, 71*(3), 26.

Gunkel, D. J. (2020). Mind the gap: Responsible robotics and the problem of responsibility. *Ethics and Information Technology, 22*(4), 307–320.

Himma, K. E., & Tavani, H. T. (2008). *The handbook of information and computer ethics*. Wiley Online Library.

Johnson, D. G. (1994). *Computer ethics*. Prentice-Hall, Inc.

Johnson, D. G. (2006). Computer systems: Moral entities but not moral agents. *Ethics and Information Technology, 8*, 195–204.

Johnson, D. G., & Noorman, M. (2013). Artefactual agency and artefactual moral agency. In *The moral status of technical artefacts* (pp. 143–158). Springer.

Johnson, D. G., & Verdicchio, M. (2023). Ethical AI is not about AI. *Communications of the ACM, 66*(2), 32–34.

Maner, W. (1996). Unique ethical problems in information technology. *Science and Engineering Ethics, 2*, 137–154.

Moor, J. H. (1985). What is computer ethics? *Metaphilosophy, 16*(4), 266–275.

Nath, R., & Sahu, V. (2020). The problem of machine ethics in artificial intelligence. *AI & Society, 35*, 103–111.

Nissenbaum, H. (2004). Privacy as contextual integrity. *Washington Law Review, 79*, 119.

Van den Hoven, J., Lokhorst, G.-J., & Van de Poel, I. (2012). Engineering and the problem of moral overload. *Science and Engineering Ethics, 18*, 143–155.
Walter, M. (1980). *Starter kit in computer ethics*. Helvetia Press and the National Information and Resource Center for Teaching Philosophy.

Chapter 2
Constituting Algorithmic Accountability

Deborah G. Johnson

> *Philosophy is thinking in slow motion. It breaks down, describes and assesses moves we ordinarily make at great speed ... It then becomes evident that alternatives are possible.*
>
> *—John Campbell.*

Abstract Although accountability is often seen as the best way to address the dangers of algorithmic decision making, too little attention has been given to understanding accountability and how it is constituted. Seeking to fill that gap, a general account of accountability is provided here and then extended to algorithmic decision making. Drawing on the work of Mark Bovens, accountability is presented as a social practice consisting of actors and forums who share a set of beliefs central to which is the belief that the actor has an obligation to explain certain kinds of behavior to a forum with the possibility that the forum may render sanctions. Importantly, the shared beliefs that constitute accountability are embedded in social norms. Extending this account to algorithmic decision making reveals a good deal of uncertainty about the actors, forums, and norms that currently constitute accountability for algorithmic decision making. Finally, when it comes to the role of algorithms in social practices of accountability, although algorithms are efficacious in algorithmic decision making, only human actors can be the actors called to account.

Keywords Accountability · Algorithmic accountability · Algorithmic decision making · AI · Social practices

---

D. G. Johnson (✉)
Department of Engineering and Society, School of Engineering, University of Virginia, Charlottesville, VA, USA
e-mail: dgj7p@virginia.edu

M. Noorman, M. Verdicchio (eds.), *Computer Ethics Across Disciplines*, Philosophy of Engineering and Technology 44,
https://doi.org/10.1007/978-3-031-87974-6_2

Introduction

Accountability is a prominent theme in media and scholarly discourse about the social implications of algorithmic decision making. This discourse grapples with the power and pitfalls of using complex computational analysis to make decisions in such domains as law enforcement and criminal justice, healthcare, banking and finance, transportation, marketing, and more. While the benefits and potential of this kind of decision making are persistently lauded, two big worries loom large, fairness (justice) and safety. That algorithms can produce racially and/or gender biased results is now well-recognized and, more broadly, there seems to be growing awareness of the non-neutrality of computational modeling and analysis. Safety concerns have been raised most conspicuously in relation to AI used in autonomous vehicles, though there are many parallel safety concerns in other contexts involving physical operations such as nuclear power plants and industrial robots.

When it comes to strategies and proposals to address these concerns, accountability is a dominant theme (Wieringa, 2020). Accountability is often at the heart of discussions of remedies that reference governance, ethics, regulation, transparency, explainability, and audits. Yet, there seems to be only superficial understanding of what accountability is. The term is frequently used as a trope for something that should happen without much thought given to what it is or what it requires. Perhaps, because so much attention has been given to AI with it being hyped as transformational and unique, the discourse on accountability frequently addresses the nature of algorithms to the neglect of the nature of accountability.

Given the centrality of the notion of accountability in the discourse on algorithmic decision making, and given how unreflectively the concept is often used, a better understanding of what constitutes accountability seems direly needed. In what follows, I provide an account of accountability to use in figuring out how to constitute algorithmic accountability. In the first part of this chapter, I present an analysis of accountability as a social practice (or set of social practices) with multiple elements. In the second part, I extend the account to algorithmic decision making. Here, I compare accountability for failures in algorithmic decision making to accountability for bridge collapses teasing out the lack of clarity about the actors, forums, and norms essential for effective algorithmic accountability practices. I end with a brief discussion of the role of algorithms in algorithmic decision making arguing that while algorithms are efficacious in algorithmic decision making, human actors are always the actors to be called to account in accountability practices.

A note about my use of the term algorithmic decision making may be helpful at the onset. I treat algorithmic decision making as an activity that always involves human actors. I am aware of the distinctions often made between humans-in-the-loop, humans-on-the-loop, and humans-out-of-the-loop systems[1] but I find these

[1] As Wieringa (2020) explains, in human-out-of-the-loop systems there is no human input or oversight; the algorithmic output is the decision. In human-on-the-loop systems, there is a default to the algorithmic output, but humans monitor the process and have an opportunity to interfere with the

distinctions misleading for accountability purposes. Conceptualizing algorithmic decision making as something that is ever without human control is a misnomer. It decontextualizes the use of algorithms and hides the fact that human actors decide when and how algorithms are used (Martin, 2019). Humans give algorithms whatever power and efficacy they have. Framing algorithmic decision making as an activity that always involves human actors more accurately depicts the way algorithms are used in real world domains. To be sure, this way of framing algorithmic decision making is much more challenging because it requires recognition of just how widely distributed such decision making can be (Neyland, 2016).

Accountability as a Social Practice[2]

To understand the notion of accountability, the work of Bovens (2007, 2014) is an important starting place.[3] Bovens provides one of the most thorough accounts of the concept (Bovens, 2007; Bovens et al., 2014). He describes accountability as "a relationship between an actor and a forum, in which the actor has an obligation to explain and to justify his or her conduct, while the forum can pose questions and pass judgement, and the actor may face consequences." (p. 450) This account is a careful specification of what Bovens believes to be a consensus view. Indeed, the elements of Bovens' account appear and reappear in theoretical discussions of accountability as well as in informal understanding and uses. In the theoretical literature, for example, Grant and Keohane (2005) specify that accountability "implies that some actors have the right to hold other actors to a set of standards, to judge whether they have fulfilled their responsibility in light of these standards, and to impose sanctions if they determine that these responsibilities have not been met." (p. 29) In another example, Schedler et al. (1999) reduces the elements of Bovens's account to just two, claiming that (political) accountability has two connotations: "answerability, the obligation of public officials to inform about and to explain what they are doing; and enforcement, the capacity of accounting agencies to impose sanctions on powerholders who have violated their public duties." (p. 14). Informally, accountability generally means that some individual is expected to explain their behavior that was at odds with the expectations of others; the others stand ready to pass judgment on the individual and render sanctions depending on the explanation provided.[4]

algorithmic output as the decision. In human-in-the-loop systems, the algorithmic output is used as a suggestion about possible actions, but no action is undertaken without human decision.

[2] This section is a revised version of Johnson (2021), pp. 111–127.

[3] I am not the first to use Bovens account to analyze algorithmic accountability. Most notably, Wieringa (2020) used it to frame a review of the literature on algorithmic accountability.

[4] This consensus view is also consistent with Adam Smith's account in *The theory of moral sentiments* (1853) where he writes: "An accountable being, as the word expresses, is a being that must

Although Boven's focus is on the accountability of political leaders and public administrative units his account is consistent with accountability in a wide variety of contexts. For example, in criminal justice, individuals (actors) are understood to have an obligation to appear in court and explain or justify their behavior when it appears to break the law. They present themselves before a judge and/or jury which represents the public (the forum). If the explanation is not satisfactory, the judge or jury may find the individual guilty and render appropriate consequences. The same essential elements are present in informal forms of accountability, for example, when a person discovers evidence of a friend's betrayal (such as revelation of an intimate secret to a third party) and demands an explanation/justification for the seeming betrayal. Friends generally share the belief that they are obligated to explain this kind of behavior. The slighted person may accept the explanation as adequate or find it lacking and, if the latter, may render consequences to the friend (for instance, breaking off the relationship). The same sort of practice is evident when a company is accused of social or environmental wrongdoing (for instance, unsustainable environmental behavior or exploitative labor practices). Here customers or citizen activist groups (forums) may call the company (the actor) to account by demanding an explanation/justification of its behavior, and if the explanation is not adequate or a change in behavior is not promised, the groups may render consequences such as boycotts or negative ratings.

Given the breadth of Bovens' account, it seems plausible to use it in thinking about algorithmic accountability (Cech, 2021; Wieringa, 2020). However, some aspects of the account need further discussion if they are to be extended to algorithmic accountability. Beginning with actors, it is worth noting that the actor or actors being held to account may be individual persons or collective entities such as corporations, government agencies, or any kind of organization. Moreover, actors may be simultaneously accountable to multiple forums. A company, for example, may be accountable (for different types of behavior) to a board of directors, regulatory agencies, employees, and customers. Here we have multiple accountability relationships.

Forums for accountability are less likely to be individuals; they are generally groups of people who believe they are entitled to an explanation. Note that on Bovens' account, forums for accountability are groups that share with the actor the belief that the actor has an obligation to account to them. Here we run into some trouble with the nuances of Bovens' account. To be sure there are many cases in which actors and forums agree. For example, in the Volkswagen emissions fraud scandal there was consensus among a wide range of forums (customers, regulatory agencies, the automotive industry, and media) as well as the company itself, that the company had an obligation to explain its apparently fraudulent behavior (Hotten, 2015). However, there are also cases in which such a belief is not shared. Typically, this occurs when forums call for an explanation and the actor does not believe they

give an account of its actions to some other, and that consequently must regulate them according to the good-liking of this other" (Sect. III).

owe the forum an explanation. Consider, for example, the case of a local citizens' group demanding information from a local manufacturing company about the chemicals being released into the local water supply. The company might refuse to provide such information claiming that it has no legal obligation to disclose such information. On Bovens' account we seem to have to say the manufacturing company is not accountable. In neither the relationship between the company and legal authorities nor the relationship between the company and the local citizens group is there a shared belief about an obligation to explain.

This doesn't seem quite right. At least from a normative perspective, we might want to insist that sometimes actors are accountable even though they deny that they have an obligation to explain their behavior. In this respect, Bovens' account does not seem to allow us to make normative claims about accountability. This is because his account is descriptive; it is a description of accountability systems or practices already in place and functioning. When accountability systems are in place, there are, as Bovens would have it, an actor and a forum that share the belief that the actor has an obligation to explain (and the forum is entitled to ask questions and render sanctions). Bovens' descriptive account does not address how accountability comes to be or how it is maintained, reinforced, and may change. Moreover, his account does not allow the possibility that accountability may be in various stages of development. For example, when a forum believes that it is entitled to an explanation but the actor it calls to account does not recognize such an obligation, the situation may be a sign of accountability 'in the making' (Johnson, 2021). The act of calling to account an actor that does not believe they owe an explanation can be seen as a first step in developing accountability, a pressuring for accountability.

Some of the difficulty here goes away if instead of making the relationship between actor and forum the centerpiece of accountability, we think of accountability as a social practice. The social practice consists of activities that involve many of the elements identified by Bovens.

Thinking in terms of social practices allows us to see that accountability is embedded in social norms and expectations. As Van Schoelandt (2018) aptly put it: "One of the reasons we seem to be able to hold each other mutually accountable as widely as we do is because our lives are infused with social norms. To a great degree, we hold each other accountable based on rules governing particular social circumstances. We guide our actions in accordance with these norms, expect the compliance of others, and hold others to those norms." (p. 223).

Social norms are patterns of belief and behavior that people adhere to, often without much thought, so they are not beliefs in the sense of strongly held convictions.[5] Indeed, individuals are often not aware of social norms that they adhere to until someone violates them. Moreover, in accountability practices it is not just the belief about the obligation to explain that may be implicit. There are social norms

[5] For more extensive discussion of social norms and a variety of issues around them including how they emerge and why and when we follow them, see Bicchieri (2005), and Bicchieri et al. (2018).

regarding when an explanation is owed (for what types of behavior), what counts as exculpatory, what are appropriate sanctions, and so on.

Of course, although many social norms are implicit (assumptions, unspoken expectations), others are explicit and formal. Arguably, the most effective means of establishing shared beliefs about when explanations are owed are formal in the sense of being specified by law, in regulations or written rules. Laws in many domains require powerful actors to explain their behavior. Think here of the requirements for public companies to disclose certain financial information. In the case of algorithmic accountability, a good example is the European Union General Data Protection Regulation (EU GDPR) which is a formal way of creating the shared belief that companies owe certain customers an explanation (articles 13–15). The EU GDPR specifies that individuals have a right to an explanation when they are turned down for benefits in processes that involve automated decision making.[6]

Nevertheless, while formal norms have advantages, informal norms can be powerful in constituting accountability practices. Remember the case, mentioned earlier, of a friend's seeming betrayal. In that type of situation, individuals have the expectation of an explanation not because there is a law but because the social norms around friendship are part of their understanding of the world, their assumptions about the meaning of friendship and how it works.

Importantly, social norms can be intentionally developed and can move from being formal to informal or vice versa. This movement is evident in discussions about changing the culture in this or that context. Changing culture means changing the prevailing social norms and beliefs about what to expect in a particular environment and can include norms for accountability. Think here of intentional and formal efforts to reduce the use of racist or homophobic behavior in some organizations. These efforts aim at changing behavior and, ultimately, changing the assumptions that individuals make about how to behave when interacting. This can lead to informal sanctions for violating the new norm.

When accountability is framed as a social practice, another important element comes into clearer view, the performative dimension. Bovens does not mention this, though it is not inconsistent with his account. The norms implicit in an accountability practice can be initiated, reinforced, and transmitted through performance. In the action of calling an actor to account, receiving their explanation, and rendering consequences, accountability is performed. Moreover, engaging in these activities even when there is no broad consensus about their appropriateness can help to establish new norms, new shared beliefs, and expectations about accountability.

Performativity helps to explain the situations mentioned earlier when an actor does not share with a forum the belief that they owe an explanation. The call by a forum for an explanation pressures the actor and encourages the belief that an explanation is owed. When, for example, a company responds to calls for corporate social responsibility by issuing reports, even if these reports fail to address the activity for

[6] The same phenomena can be explained in multiple ways so there is a complex issue about what is an appropriate or meaningful explanation (Selbst & Powles, 2018); it would seem to depend on the forum for the explanation.

which they are being called to account, their reporting performs accountability. Suppose that in response to public calls for corporate social responsibility, a corporation reports on its charitable activities instead of its degradation of the environment, in responding to the call, the corporation, acknowledges that it has social responsibility. It performs and, as such, reinforces the idea that a public forum is entitled to an explanation. Performances of this kind help to create accountability practices where they are not firmly established.

So far, little has been said about sanctions and beliefs about forums imposing them. This important aspect of accountability practices cannot be fully understood without addressing the function of accountability. Having recognized that accountability is a social practice, we now can ask, why do such practices exist? What purpose do they serve? Accountability is generally understood to be directed at constraining power. It puts constraints on the ability of actors to exercise their power. Schedler et al. (1999) explained how this works: "Rather than denoting one specific technique of domesticating power, it [accountability] embraces three different ways of preventing and redressing the abuse of political power. It implies subjecting power to the threat of sanctions; obliging it to be exercised in transparent ways; and forcing it to justify its acts." (p. 14).

Notice that Schedler sees the threat of sanctions as both 'preventing and redressing' the abuse of power. Here we see how accountability is both forward- and backward-looking. Accountability practices can deter actors from abusing their power as well as respond to those who have not been deterred. These two functions work together in the sense that when the backward-looking practice is performed, e.g., an actor is held to account for their bad behavior, this reinforces the threat of sanctions for that behavior and thereby may deter others from engaging in that kind of behavior in the future.

As we turn to algorithmic accountability, it is worth mentioning that the mere threat of formal accountability practices can pressure actors to create informal norms for accountability. For example, industries sometimes adopt self-regulatory practices and disseminate best practices to avoid regulation. A statement by Dolshi-Velez et al. (2017) suggests that the threat of regulatory control of AI is pressuring some in the field to develop informal norms. Dolshi-Velez et al. write that the "question of how to hold AI systems accountable is important and subtle: poor choices may result in regulation that not only fails to truly improve accountability but also stifles the many beneficial applications of AI systems." (p. 2).

So, accountability is a social practice consisting of actors and forums sharing a set of beliefs central to which is the shared belief that the actor has an obligation to explain certain kinds of behavior and may be liable to sanctions from the forum. These shared beliefs are embedded in social norms that may be formal and/or informal. The social practice is directed at putting a check on the exercise of power.

Algorithmic Accountability

We can now use this account to understand algorithmic accountability. Returning to the two concerns mentioned at the very beginning of this chapter—fairness and safety, we can think of algorithmic accountability as directed at constraining the power of actors who use algorithms to make decisions involving safety and fairness. Much of the current discourse suggests that algorithmic accountability is not well-established, that is, many scholars and journalists seek strategies to protect against unfair and dangerous uses of algorithms. This is especially so for domains in which algorithms, especially AI algorithms, have only recently been introduced, or are anticipated to be introduced. Optimistically, we can think of algorithmic accountability as 'in the making' (Johnson, 2021), that is, we can view the current discourse on algorithmic accountability as representing negotiations about how to constitute effective accountability practices.

Framing algorithmic accountability as a social practice directs us to focus on the relevant actors to call to account, the forums to which these actors should be expected to account, the beliefs and norms that should constitute the callings to account, when where and how this is done (performed), and the sanctions that can be rendered. The primary focus in what follows will be on the actors, forums, and norms in algorithmic decision making.

Actors, Forums, Norms, and the Many Hands in Algorithmic Decision Making

Algorithmic decision making typically takes place in contexts in which there is a complex division of labor with multiple individuals and organizational units and subunits ('many hands') contributing to a decision.[7] Importantly, each of the many actors may be part of a different accountability practice, and each practice may be constituted by norms specific to the individual or unit, e.g., beliefs about the goals of that unit, prevailing standards for that activity, organizational culture and values, regulatory requirements.[8]

Of course, algorithmic decision making is not unique in being constituted by many hands. Comparison to another technological context in which accountability practices are well-established may be helpful here. Consider what happens when a

[7] Helen Nissenbaum is, to my knowledge, the first to explicitly use the 'many hands' concept to explore accountability in computerized decision making (Nissenbaum, 1996).

[8] Recognizing the many hands involved, scholars have suggested new units of analysis to acknowledge this. Percy et al. (2021) suggests that we think of ecosystems when it comes to AI accountability. Ananny (2016) has coined the term 'networked information algorithms' (NIAs). Metcalf et al. (2022) refer to accountability regimes. Many others use the notion of assemblages.

bridge collapses.[9] Calls for accountability are immediate. They come from a variety of forums and are made to the key actors responsible for the bridge. The number of actors called to account may expand as more comes to be known about what happened, but initially at least, the first-line actor to be called to account will be a government agency. This is because most bridges are built under the auspices of a local, state, or federal agency. In most bridge collapses, attention then turns to those who built the bridge. Either way, the first-line actors in bridge collapses generally accept without question that they are obligated to provide an explanation of what happened.

Although many specialized forums may be activated, the broadest forum understood to be entitled to an explanation is 'the public'. The public is largely represented by and through legal authorities and media. The public expects to receive an explanation of why the bridge collapsed, who or what was at fault, and what will be done to prevent such collapses in the future.

An investigation into the cause of the bridge collapse will then take place, usually by a team of engineers and other experts. Although we generally think of the investigation as an investigation into the cause of the collapse, it is important to note that identification of the cause depends heavily on norms and standards for relevant actors. For example, were the investigative team to conclude that the bridge collapse was caused by the failure of a particular beam, there would be further investigation as to why the beam failed. Was it the material used in the beam? Was the beam properly installed? Was too much weight put on the bridge? Should maintenance people have detected a problem with the bridge before it collapsed? and so on. Standards and norms are essential to answer these questions. What counts as a good beam? What are the standard methods for calculating the size of beams to be put in certain places? What are the proper methods for installing beams? Was the amount of pressure put on the beam more than specified by the designers? And so on. The questions about cause and who is at fault can only be answered by referring to norms for bridge building and beams.

So, accountability in bridge collapses involves many actors and each actor operates with a set of beliefs and norms about how they are to operate to ensure the safety of the bridge. Each actor may be held to account by a different forum or forums. Yes, all are ultimately accountable to the public, but in the process of accounting for a bridge collapse, many actors may be simultaneously accountable to forums within their own organizations (e.g., to company executives, to an organizational unit, to colleagues) as well as to outside forums such as professional associations and industrial standard setting organizations. In short, accountability for bridge collapses starts with calling first-line actors to account but this inevitably leads to calling many other actors to account. Each actor held to account will be expected to explain their behavior in terms of a set of norms for actors in that role.

What does algorithmic accountability look like by comparison? Of course, bridge collapses are failures, so to make the two cases parallel, failures in algorithmic decision making should be the focus. What is failure in algorithmic decision

[9] See Schiaffonati and Verdicchio (2018) for a comparison of bridges and software.

making? This is not a simple question, but for our purposes here, we can consider as failures, algorithmic decision making that is unsafe or unfair. For example, when an autonomous vehicle relying on AI algorithms hits a pedestrian in its path or when a company using an algorithmic recruiting system discriminates against applicants who are female or non-white.

Actors

Bridges and algorithmic decision making are alike insofar their production involves a multiplicity of actors. On the other hand, with bridges, when there is a failure, the first-line actors to be called to account are easy to identify. It helps that bridges are usually public endeavors and that their failures are visible to the public, dramatically so. Algorithmic decision making can be a public endeavor (e.g., done by a government agency) or a private undertaking and sometimes a mixture. Either way, algorithmic decision making is generally hidden from public scrutiny. Importantly, there is currently a good deal of uncertainty about who the first-line actors are or should be when algorithmic systems are found to be unsafe or unfair, that is, it is not always clear who has an obligation to explain when a failure in algorithmic decision making comes to light. This uncertainty is consistent with algorithmic accountability being in the early stages of development. To be sure, in some domains there seems to be more consensus than others. For example, when accidents have occurred with Tesla automobiles, the U.S. National Highway Traffic Safety Administration (NHTSA) has turned to Tesla as the first-line actor to be held to account (Blanco, 2023). Similarly, when Microsoft released its AI chatbot, Tay, Microsoft was the obvious actor to be held to account for the spread of racist jokes (Wolf et al., 2017). In other cases, such as when a recruiting system fails, it is unclear who is the first-line actor to be held to account and who should lead an investigation of the many hands involved.

When there is this kind of uncertainty about first-line actors, the discovery of a failure often leads to finger pointing in many different directions with many actors deflecting an obligation to explain. Even in Tesla accidents, Tesla has claimed that the drivers were at fault because the owner's manual specified that drivers should attend to driving even when the car is in self-driving mode (Autoweek Editors, 2018). Crawford and Schultz (2019) explore the possibility of holding AI vendors accountable for the systems they provide to government agencies precisely because the government agencies that use AI systems claim not to understand how the AI systems work and, for that reason, deny having an obligation to explain.

Again, optimistically we can think of the finger pointing and deflection here as part of public negotiations over who ultimately should be the first-line actors in algorithmic decision making failures. Not surprisingly, some of the negotiations are being worked out in courts of law, especially in tort law. Here we see that users (usually organizational entities) that use algorithmic systems to make decisions, are often the actors held to account in suits by those harmed by their decisions.[10] Given that accountability practices are mechanisms for constraining the exercise of power, focusing on users is not an unreasonable locus for first-line accountability insofar as holding users to account pressures them to avoid using algorithms unless they have been shown to be safe and/or fair.

However, my point here is not to push for users, or any other specific category of actors, to be the first-line actors, rather my point is that the first-line obligation to explain should be assigned to the locus (the actors) where it will do the most good. Since accountability practices are mechanisms for constraining power, this is a matter of identifying and assigning the first-line obligation to explain to the actors who have the most power to prevent unsafe and unfair decision making and/or those who are in the best position to call others to account.

Importantly, this is not just a matter of figuring out where power currently lies for this just takes us down the path of seeing the many hands and the disparate and distributed power in algorithmic decision making. Instead, the focus should be on effectiveness even if this involves rearranging or creating power. For example, making algorithm users the first-line actors in accountability may shift some power away from algorithm designers or vendors by putting the burden of establishing standards on the users and making designers more vulnerable to the demands of users. Similarly, making the vendors of algorithmic based systems the first-line actors may give them more power to set standards and be less responsive to user concerns. In any case, assigning the first-line obligation to explain to one set of actors will not let any other actors off the hook. The first-line actors, as with bridge collapses, can, in turn, call the other actors to account.

The bottom line here is that because of the many hands involved in algorithmic decision making, accountability seems now to consist of an endless game of finger-pointing in multiple directions with the result that no one accepts responsibility for the entire enterprise of decision making. Instead, the focus should be on constituting accountability practices such that when failures occur, first-line actors are easy to identify, have the authority to investigate what happened, and acknowledge the obligation to explain. The forward- and backward-looking valences of accountability are important to remember here for assigning the first-line obligation to explain to a particular category of actors will simultaneously assign those actors a duty to do what they can to prevent failures from occurring.

[10] Fraser et al. (2022) suggest that tort law be used while waiting for stronger forms of protection for those who are harmed by algorithmic decision making.

Forums

Actors and forums go hand in hand; holding actors to account requires forums to call for, receive, and evaluate explanations. Bridge collapses and algorithmic decision making are alike in requiring multiple forums for the many different actors involved, but again here, unlike bridge collapses, there seems to be a good deal of uncertainty about the appropriate forums for algorithmic decision making. Media coverage suggests that there is public interest in holding actors to account but the belief that any forum (other than the law) is entitled to an explanation is not deeply inculcated in public expectations.[11] In addition to law and the media, the public is now represented by a variety of non-profit organizations such as AI Now and CAIDP. As well, there are professional communities and organizations, groups of academic scholars and experts including many computer scientists who have become forums through their public discourse on algorithmic decision making.

Because most algorithmic decision making is embedded in institutional contexts, the forums for accountability practices tend to vary from one domain to another, e.g. for transportation, financial, and healthcare systems. This is especially so when it comes to formal forums which may have existed prior to the use of algorithms and already do or could or should serve as the appropriate forums for algorithmic decision making. For example, the U.S. National Transportation Safety Board (NTSB) currently serves as a forum for holding actors in the transportation industry to account and they have become a formal forum for self-driving vehicles. The Federal Reserve System serves as a forum for the decisions made by many financial institutions, and they, too, can hold financial institutions to account for their use of algorithms. But there are many other domains in which there are no formal forums for accountability.

Formal forums aside, in many domains the opacity of decision making (algorithmic and non-algorithmic) is legally protected from public scrutiny by proprietary, trade secrecy, and competitive practices law. Cech (2021) suggests that we think about forums in terms of their agency; if forums are to have agency they must be "aware of (1) the existence of the system, (2) the identity of the actor (e.g., the developers, maintainers, or product owners of the system) and (3) have means of communicating their inquiries. This directs Cech to the importance of transparency. How can forums such as watchdog and citizen action organizations, journalists, legal authorities, or any group of individuals call actors to account when these forums are in the dark about what the actors do. The lack of transparency is, indeed, an obstacle to the work of forums and in this way, an obstacle to algorithmic accountability (de Laat, 2020).

However, for transparency to be effective in algorithmic accountability practices, much more attention needs to be given to establishing effective forums, especially forums that make a strong claim that they are entitled to explanations of algorithmic

[11] Metcalf et al. (2022) express concern that the public will be bypassed in some algorithmic accountability regimes.

decision making. Establishing effective forums is not a simple matter. For example, the GDPR establishes a legal right to an explanation for those who are turned down for a benefit by algorithmic decision making.[12] Although this is an important right, the GDPR gives this right to individuals (who have been harmed by an algorithmic decision) and not to the public (Goodman & Flaxman, 2017). At best, it is unclear whether this is an effective check on the power of those who make decisions that significantly affect the lives of individuals. At worst, this strategy diffuses the possibility of a collective forum with the capacity to call for a public explanation.[13]

Norms

As we saw in the case of bridge collapses, norms are essential for any forum to hold actors to account. Accountability practices rely on norms with regard to the kinds of behavior that need to be accounted for,[14] when, where and how the accounting is to be done, what will count as sanctionable, and what sanctions can be rendered, and more. While norms for bridge building, including norms for beams, are well-established, one of the most striking obstacles to algorithmic accountability is the lack of norms for many of the actors involved in algorithmic decision making, from the designers and users of algorithms, and for many other activities that contribute to algorithmic decision making, e.g., the training of algorithmic designers, the vendors of algorithmic software, the organizational culture of the institutions using algorithms, to the training of users. Even if we focus just on the design of algorithms, as we might with bridge beams, there are multiple crucial decision points in algorithmic design and very few norms for making these decisions. Cummings (2021) characterizes the decision points as subjective, pointing to a lack of norms when it comes to how algorithm designers select data sets, modify the data sets to make them compatible, decide when a model is accurate, choose which features of the output are significant, and assign weights to features.

Again, transparency comes into play here because transparency about decisions in algorithmic design is necessary if designers are to be held to account. However, transparency will not be effective unless there are norms to evaluate what is made transparent. Norms are required to establish when, where, for whom, and what should be disclosed (Fung et al., 2007). Diakopoulos (2014) discusses standardizing what should be disclosed about algorithms (in the context of journalism). Focusing just on the algorithm, he suggests that standards might reference:

---

[12] See https://gdpr.eu/tag/gdpr/

[13] Kemper and Kolkman (2019) emphasize the need for a critical audience for algorithmic accountability.

[14] Donovan et al. (2018) point out that there are not standard definitions for what counts as algorithmic bias.

> (1) the criteria used to prioritize, rank, emphasize, or editorialize things in the algorithm, including their definitions, operationalizations, and possibly even alternatives; (2) what data act as inputs to the algorithm—what it "pays attention" to, and what other parameters are used to initiate the algorithm; (3) the false positive and false negative rate of errors made in classification, including the rationale for how the balance point is set between those errors; (4) training data and its potential bias, including the evolution and dynamics of the algorithm as it learns from data; and (5) the definitions, operationalizations, or thresholds used by similarity or classification algorithms (p. 2).

Yes, there are some informal norms for some of the decision points involving algorithm design and use, for example, norms are transmitted when students are trained in how to develop algorithms and when to consider them valid. There are implicit norms in published research material as well as in various professional activities. Nevertheless, such norms, to the extent that they exist at all, are not well-articulated or -promulgated; they do not seem to be in a form that can be used for public accountability. The lack of norms may in part explain why so much attention has been given to creating principles (de Laat, 2020; Percy et al., 2021). Principles are norms though, like codes of professional ethics, they are norms at such an abstract level that they require interpretation to have any meaning.[15]

The Role of Algorithms in Algorithmic Accountability

As just mentioned, algorithms seem parallel to beams in the bridge collapse case. Just as we might identify a beam as the locus of failure in a bridge collapse, we might identify an algorithm as the locus of failure in algorithmic decision making.[16] Of course, identifying the beam as the locus of failure does not end the accountability story nor should identifying an algorithm as the source of failure in algorithmic decision making. Locating the algorithm as the source (the cause) of a problem in a decision-making system is merely the starting place for a whole host of questions. Accountability depends on the prevailing norms for selecting, using, designing, testing, and training of algorithms. So far, my discussion of accountability as a social practice has said little about the role of algorithms except for the apparent lack of norms around their design and use.

This lack of attention may seem odd since so much of the literature on algorithmic accountability focuses on the nature of algorithms. One way to bring algorithms into the account would be to consider them as actors, actors that forums can call to account, demanding an explanation of their behavior. This is suggested by those who put the emphasis on developing explainable algorithms.

---

[15] See Johnson (2021) for a discussion of the abstractness of professional codes of ethics.

[16] Diakopolous (2015) identifies the types of processes that algorithms perform as prioritization, classification, association, and filtering.

An argument can be made for thinking in this way by drawing on the literature on the agency of technology, especially Actor Network Theory (ANT).[17] In ANT, technological systems are understood to be networks comprised of human and non-human nodes that together produce outcomes. So as not to treat human and nonhuman components differently, all nodes in the network are referred to as actants (Latour, 1996; Law et al., 2012). The virtue of thinking in terms of actants is that it brings into focus how efficacious non-human entities, such as algorithms, are in producing outcomes. This helps us to see that certain outcomes might not have been achieved without the non-human entities. For example, in the case of a decision made using an algorithm, the decision might not have been achievable without the algorithm, or at least not in the form made possible by the algorithm. Algorithms (and computation in general) make pragmatically possible certain kinds of precision and scale not possible without them. Hence, thinking of algorithms and other non-human entities as efficacious agents in producing outcomes is important and useful.

Nevertheless, when it comes to accountability, the distinction between agency and actors is important. Remember that the function of accountability is to put constraints on the exercise of power. Accountability achieves this function by instilling in both actors and forums the belief that the actor has an obligation to explain certain types of behavior and that the actor will face consequences (sanctions) if the explanation is not forthcoming or adequate. As a mechanism for constraining power, accountability practices do not work on algorithms. Threats of sanctions cannot motivate an algorithm; rather threats of sanctions motivate the human actors involved in the production and use of algorithms, be these individuals or organizations. Motivated designers can design (or at least try to design) their algorithms so as not to have harmful effects, and users can be motivated to use algorithms only in ways that are not harmful. Accountability can even motivate algorithm designers to make their algorithms transparent or to make them produce explanations of their operations (Doshi-Velez et al., 2017). In this respect, algorithms can be designed in ways that facilitate the social practice of calling human actors to account.

The other, perhaps obvious, reason for not thinking of algorithms as actors in accountability practices is that algorithms are not the kind of entities that have obligations and beliefs. Yes, we can talk 'as if' they can be held accountable; for example, we could say that when algorithms are designed to give explanations of their operations, they become 'obligated' to explain their behavior. This metaphorical way of thinking may be useful for certain purposes but what it means for a human to have an obligation is quite different from what it means to speak as if algorithms have obligations.

This distinction between actors and entities that are efficacious in producing outcomes will likely be challenged by those who imagine computational systems of the future that have general intelligence comparable to that of human beings. Whatever might be said about future algorithms, it is unclear just how the threat of sanctions

[17] Also, in legal contexts, some discussion has begun about giving algorithms status as artificial persons. See, Barnett et al. (2017).

would work on such entities especially since embodiment is relevant to sanctions. The threat of sanctions is generally a threat to an actor's physical or financial well-being. In any case, however computers may operate in the future, for now the function of accountability cannot be achieved by treating algorithms as actors in social practices of accountability.

Thinking about the role of algorithms in accountability practices leads us to an important distinction between entities that can be held to account, entities that can be accounted for, and entities that can shape practices of accountability. These are not mutually exclusive categories; individuals and organizations can be classified in all three categories. Algorithms are typically entities to be accounted for because they are part of decisions or events for which an explanation may be owed. Algorithms are also entities that shape the character of accountability practices. For example, when parole boards shift to using algorithmic systems to predict recidivists, the parole board's accountability for its decisions shifts to include its use of the algorithmic system. The use of algorithms may also shape accountability systems by affecting how organizations keep track of and report on their own behavior (Johnson & Regan, 2014).

So, algorithms are important components in algorithmic decision making but their role is not that of actors to be called to account. Rather, they are entities for which humans are accountable. Humans are accountable for how algorithms operate and for how they are used in decision making. The many human hands that are involved in algorithmic decision making can be motivated by accountability practices to constrain the power of the algorithms they use.

Conclusions

This chapter began with the idea that in the current discourse on algorithmic accountability, too little attention is paid to the nature of accountability. The account of accountability provided here—a modified version of Bovens' account—suggests that the social practices that might constitute algorithmic accountability are not well-established. The account identifies what is needed to more fully establish algorithmic accountability. This includes assigning first-line actors the duty to explain when harm is done as a result of algorithmic decision making. It includes inculcating the belief that public forums are entitled to explanations when failures occur and harm is done. As well, it requires establishing norms for all the activities around algorithm design and use. Little has been said about when and how algorithmic accountability can or should be performed and what sanctions can or should be used, but these matters will follow when more clarity is achieved about actors, forums, and norms.

The many hands involved in algorithmic decision making makes the challenge of constituting effective accountability practices daunting. On the other hand, the good news is that efforts to establish algorithmic accountability are likely to improve accountability for decision making more broadly, i.e., without algorithms. When individuals and organizations expect to be called upon to explain their behavior, they are compelled to think more carefully about it, whether that behavior involves relying upon algorithms or anything else. This is one of the themes that has run through the history of computing. To make use of the power of computation, people are compelled to make their goals and strategies explicit so that they can be put into the form of a program or algorithm. The articulation process can be powerful in itself, even before automation takes place. Unfortunately, when the process is hidden or left to programmers who don't fully understand the activity and what may be at stake in tweaking the algorithm this way or that, short-cuts are likely to be taken, leading to dangerous and unfair results.

What is clear from the preceding analysis is that explanations of how algorithms work, e.g., explainable AI, is not the be all and end all of accountability. Algorithms do have to be accounted for but accounting for them requires forums that call human actors to account for their behavior and the behavior of the algorithms, and that, in turn, requires norms that allow actors to understand what they should and shouldn't do.

References

Ananny, M. (2016). Toward an ethics of algorithms: Convening, observation, probability, and timeliness. *Science, Technology, & Human Values, 41*(1), 93–117.

Autoweek Editors. (2018, April).Tesla continues to blame driver in fatal crash involving autopilot, *Autoweek.* https://www.autoweek.com/news/technology/a1695306/tesla-blames-driver-fatal-crash-involving-autopilot/

Barnett, J., Koshiyama, A. S., & Treleaven, P. (2017). Algorithms and the law. Retrieved from Legal Futures: https://www.legalfutures.co.uk/blog/algorithms-and-the-law

Bicchieri, C. (2005). *The grammar of society: The nature and dynamics of social norms.* Cambridge University Press.

Bicchieri, C., Muldoon, R., & Sontuoso, A. (2018). *Social norms, the Stanford encyclopedia of philosophy* (Winter 2018 Edition). https://plato.stanford.edu/archives/win2018/entries/social-norms/

Blanco, S. (2023, June 13). Report: Tesla autopilot involved in 736 crashes since 2019, car and driver. https://www.caranddriver.com/news/a44185487/report-tesla-autopilot-crashes-since-2019/

Bovens, M. (2007). Analysing and assessing accountability: A conceptual framework. *European Law Journal, 13*(4), 447–468.

Bovens, M., Goodin, R. E., & Schillemans, T. (Eds.). (2014). *The Oxford handbook public accountability.* Oxford handbooks.

Cech, F. (2021). The agency of the forum: Mechanisms for algorithmic accountability through the lens of agency. *Journal of Responsible Technology, 7*, 100015.

Crawford, K., & Schultz, J. (2019). AI systems as state actors. *Columbia Law Review, 119*(7), 1941–1972.

Cummings, M. (2021). Rethinking the maturity of artificial intelligence in safety-critical settings. *AI Magazine, 42*(1), 6–15.

de Laat, P. B. (2020). From algorithmic transparency to algorithmic accountability? principles for responsible AI scrutinized. *ETHICOMP, 2020*, 336.

Diakopoulos, N. (2014). Algorithmic accountability reporting: On the investigation of black boxes. academiccommons.columbia.edu

Diakopoulos, N. (2015). Algorithmic Accountability, Digital Journalism, 3(3), 398–415. https://doi.org/10.1080/21670811.2014.976411
Donovan, J., Caplan, R., Matthews, J., & Hanson, L. (2018). Algorithmic accountability: A primer.
Doshi-Velez, F., Kortz, M., Budish, R., Bavitz, C., Gershman, S., O'Brien, D., … Wood, A. (2017). Accountability of AI under the law: The role of explanation. *arXiv preprint arXiv:1711.01134.*
EU GDPR articles 13–15, https://gdpr.eu/tag/gdpr/.
Fraser, H., Simcock, R., & Snoswell, A. J. (2022, June). AI opacity and explainability in tort litigation. In *Proceedings of the 2022 ACM conference on fairness, accountability, and transparency* (pp. 185–196).
Fung, A., Graham, M., & Weil, D. (2007). *Full disclosure: The perils and promise of transparency.* Cambridge University Press.
Goodman, B., & Flaxman, S. (2017). European Union regulations on algorithmic decision-making and a "right to explanation". *AI Magazine, 38*(3), 50–57.
Grant, R. W., & Keohane, R. O. (2005). Accountability and abuses of power in world politics. *American Political Science Review, 99*(1), 29–43.
Hotten, R. (2015). Volkswagen: The scandal explained. *BBC news, 10*, 12. https://www.bbc.com/news/business-34324772
Johnson, D. G. (2021). Algorithmic accountability in the making. *Social Philosophy and Policy, 38*(2), 111–127.
Johnson, D. G., & Regan, P. M. (Eds.). (2014). *Transparency and surveillance as sociotechnical accountability: A house of mirrors*. Routledge.
Kemper, J., & Kolkman, D. (2019). Transparent to whom? No algorithmic accountability without a critical audience. *Information, Communication & Society, 22*(14), 2081–2096.
Latour, B. (1996). On actor-network theory: A few clarifications. *Soziale Welt*, 369–381.
Law, J., Bijker, W. E., Hughes, T. P., & Pinch, T. (2012). Technology and heterogeneous engineering: The case of Portuguese expansion. *The Social Construction of Technological Systems: New Directions in the Sociology and History of Technology, 1*, 105–128.
Martin, K. (2019). Ethical implications and accountability of algorithms. *Journal of Business Ethics, 160*, 835–850.
Metcalf, J., Moss, E., Singh, R., Tafese, E., & Watkins, E. A. (2022). A relationship and not a thing: A relational approach to algorithmic accountability and assessment documentation. *arXiv preprint arXiv:2203.01455.*
Neyland, D. (2016). Bearing account-able witness to the ethical algorithmic system. *Science, Technology, & Human Values, 41*(1), 50–76.
Nissenbaum, H. (1996). Accountability in a computerized society. *Science and Engineering Ethics, 2*, 25–42.
Percy, C., Dragicevic, S., Sarkar, S., & d'Avila Garcez, A. (2021). Accountability in AI: From principles to industry-specific accreditation. *AI Communications, 34*(3), 181–196.
Schedler, A., Diamond, L. J., & Plattner, M. F. (Eds.). (1999). *The self-restraining state: Power and accountability in new democracies*. Lynne Rienner Publishers.
Schiaffonati, V., & Verdicchio, M. (2018). What do bridges and software tell us about the philosophy of engineering? *Philosophy of Engineering, East and West*, 137–148.
Selbst, A., & Powles, J. (2018, January). "Meaningful information" and the right to explanation. In *Conference on fairness, accountability and transparency* (pp. 48–48). PMLR.
Smith, A. (1853). *The theory of moral sentiments*. HG Bohn.
Van Schoelandt, C. (2018). Moral accountability and social norms. *Social Philosophy and Policy, 35*(1), 217–236.
Wieringa, M. (2020, January). What to account for when accounting for algorithms: A systematic literature review on algorithmic accountability. In *Proceedings of the 2020 conference on fairness, accountability, and transparency* (pp. 1–18).
Wolf, M. J., Miller, K., & Grodzinsky, F. S. (2017). Why we should have seen that coming: comments on Microsoft's tay "experiment," and wider implications. *Acm Sigcas Computers and Society, 47*(3), 54–64.

Chapter 3
Algorithmic Accountability Gets the Deborah Johnson Treatment

Keith W. Miller

Abstract A central task of computer ethics is to analyze problems that can be described as technical details intertwined with human values. Traditionally, computer professionals (like other practitioners in engineering fields) have mistrusted philosophers, whose thinking they characterize as too abstract, and not sufficiently grounded in the reality of computing. Some philosophers suggest that computing professionals (and other engineers) over-emphasize the importance of technical issues while minimizing the importance of ethically significant principles. The genius of Deborah Johnson has been to engage with both computing professionals and philosophers and to develop insights that help increase understanding among both groups. This chapter explores how her recent work on algorithmic accountability illustrates Johnson's approach, and how she succeeds in clarifying several crucial issues.

Keywords Algorithmic accountability · Deborah Johnson · Social practices

Introduction

The title of this chapter requires an explanation. The use of the phrase "the Deborah Johnson treatment" is not to suggest that Johnson deploys a one-size-fits-all recipe for churning out ideas about computer ethics. Much the opposite, I will argue that Johnson's "treatment" includes rigorous scholarship, a long view both forward and backward, and attention to detail that illuminates the challenges. When you work with computers, attention to detail is a crucial skill.

K. W. Miller (✉)
College of Education and Department of Computer Science, University of Missouri–St. Louis, Chatham, IL, USA
e-mail: millerkei@umsl.edu

M. Noorman, M. Verdicchio (eds.), *Computer Ethics Across Disciplines*, Philosophy of Engineering and Technology 44,
https://doi.org/10.1007/978-3-031-87974-6_3

I contend that Johnson's habits of the heart (a phrase made famous by Robert Bellah (Bellah et al., 1996), and attributed to Alexis De Tocqueville (2015)) lead her to focus on the aspects of an issue that matter most to society, and ultimately to human beings. Johnson's passion about justice and human thriving, combined with her meticulous, patient scholarship, are why I think she is such an important figure in computer ethics.

Johnson's (2021) article "Algorithmic accountability *in the making*" (author's emphasis) is a good example of her approach. This work starts with a contemporary concern (in this case, calls for accountability around algorithms, including AI algorithms), and weaves ideas and details in a way that is clear, but sophisticated. The article points out that much of the discussion of algorithmic accountability focuses on the algorithms themselves (particularly AI algorithms' lack of transparency), and Johnson cautions against making that the exclusive area of concern. Instead, she insists that accountability is best understood as a social practice, a practice that ultimately can constrain the exercise of power by organizations and individuals. As we shall see, this emphasis is rooted in Johnson's earlier work, and then applied to important new technical details of more recent technological developments.

Johnson's Past Work on Accountability Foreshadows Her Recent Work on Algorithmic Accountability

In this section, I will discuss three of Johnson's previous works that predict her recent ideas about algorithmic accountability.

1995 Accountability and Computer Decision Systems

Public interest and research articles about the ethics of artificial intelligence (AI) have increased explosively in the past few years (Borenstein et al., 2021). But some scholars have been working on associated questions for decades. In 1995, Johnson and Mulvey (1995) published "Accountability and computer decision systems." That article does not mention artificial intelligence, but does discuss in detail "computer decision systems," that accomplish tasks that AI applications are being used for today. Several of Johnson and Mulvey's suggestions about accountability for those systems in 1995 presage arguments she has made about algorithmic accountability more recently.

This paper appeared in *Communications of the ACM,* the flagship publication of a leading organization for computing professionals. It speaks both to serious philosophical questions and to practical challenges of engineers. Johnson and Mulvey describe the close relationship between accountability and responsibility, pointing to Michael Davis's work in professional ethics (Davis, 1991). Johnson and Mulvey focus primarily on the social context around computer decision systems, a theme we see reprised in Johnson's work on algorithmic accountability.

2007 Ethics and Technology "In the Making": An Essay on the Challenge of Nanoethics

The title of Johnson's 2021 paper on algorithmic accountability includes the phrase "in the making." That phrase also appears in a 2007 paper entitled "Ethics and technology 'in the making': an essay on the challenge of nanoethics," published in the inaugural issue of the journal *Nanoethics* (Johnson, 2007). This paper includes two themes which appear prominently in much of Johnson's work: the rejection of technological determinism, and the embrace of sociotechnical systems as the proper conceptualization of technology. Both these themes are important in the field of science and technology studies (STS) (Sismondo, 2010), a literature on which Johnson relies, and to which she contributes.

By insisting on examining sociotechnical systems, STS consistently draws us to the people involved in the design, development, and deployment of technology, while pulling us away from an exclusive focus on the technical details of artifacts. STS does not mask the importance of technical details but requires that such details always be viewed in their proper context.

Johnson and STS consistently remind us that because humans are a vital part of all sociotechnical systems, humans can (and should) be in control of the development of these systems. This opportunity is also a responsibility, and the accountability for algorithms is deeply linked to the accountability for nanotechnology, and AI.

Although the phrase "in the making" seems particularly appropriate when technologies are first emerging, it seems clear in Johnson's work that many technologies are in some sense always "emerging." Sociotechnical systems are constantly being adapted and adopted, finding new ways to affect people. As such, we should never ignore the possibility that we (as individuals and as a society) can initiate changes to these systems and how they are used. Johnson's application of STS principles is at once a call to engage with our ethical responsibilities, as well as an invitation to be optimistic that real change is still possible when we do. An emphasis on the emerging nature of all sociotechnical systems helps us avoid the dangers of ethicists becoming ambulance-chasers, focusing only on systems that happen to have caught the public's attention now.

2015 Technology with No Human Responsibility?

This 2015 paper appeared in *The Journal of Business Ethics* and is another example of how Johnson has been successful at engaging with broad, inter-disciplinary audiences. It also develops arguments later refined for use in the 2021 paper on algorithmic accountability.

An important question is whether so-called "artificial agents" can ever be thought of as responsible for their own behavior, thereby relieving the humans who design, develop, and deploy those artifacts of responsibility. This is a question that Johnson has discussed often, both before and after 2015. For example, see Johnson and Powers (2005), Johnson (2006), Johnson and Miller (2008), Johnson and Noorman (2014), and Johnson and Verdicchio (2019). (This list is not exhaustive.)

Though the details of Johnson's arguments evolve through the years, her fundamental answer to the question is consistent: No, artificial agents (no matter how sophisticated) should not be considered morally responsible for their own behavior; humans are responsible for those "agents." If an artificial system is unpredictable and harmful (perhaps because it is designed to "learn"), then the humans who designed, developed, and deployed that system should be accountable.

In the 2015 paper, Johnson argues against "the responsibility gap," the notion that AI that learns creates a new situation, where those who make the technology cannot predict its future behavior, and then (some contend) the developers cannot be ascribed traditional responsibility. Andreas Matthias wrote, "…society must decide between not using this kind of machine anymore (which is not a realistic option), or facing a responsibility gap, which cannot be bridged by traditional concepts of responsibility ascription." (Matthias, 2004, p. 175).

Johnson describes several reactions to the responsibility gap. Following Santoro et al. (2008), she points out that society already demands, in certain cases, responsibility without explicit fault. The paradigm example is the legal principle of strict liability. We note here that strict liability is a method of social control that asserts, in effect, that if you decide to profit from this product (despite its inherent dangers), then if those dangers manifest, you will be accountable (in this case, legal and financially accountable). Society could decide that AI artifacts that are unpredictable are sufficiently analogous to a dangerous product that those who profit from them should be held to strict liability.

Strict liability is only one possible response to claims that learning AI "breaks the rules" of traditional professional ethics for engineers. Any such response, Johnson points out, will require society to make conscious decisions, and codify these ideas in regulations, laws, and practices. Indeed, at this writing, several examples are emerging of legislation and proposals meant to reign in AI applications (Chamberlain, 2023; Hacker et al., 2023).

Another response to the supposed responsibility gap is to work towards making AI applications that will behave ethically. The hope is that this avoids unfairly burdening AI developers, and instead avoids ethical breaches by "programming in" ethical behavior. One prominent advocate for this position, especially in military applications, is Ron Arkin (2009).

A third response, less well known in 2015 than it is today, is to claim that at some time in the future, AI applications will become sufficiently advanced that society will ascribe to them a status that will include responsibility for their own actions, separate from the people who deployed the applications (Hellström, 2013).

When this article was published, envisioning highly autonomous robots required imaginative vision, supposedly far into the future. Today, the progress of AI applications makes it easier to envision that future. This makes even more relevant Johnson's contention that it is dangerous to assume the inevitability of autonomous robots. If society resigns itself to such machines (even before they exist), then society is unlikely to seriously consider restricting such machines because of the potential dangers. Surely one way to effectively deal with a responsibility gap in machines is to avoid such machines. Johnson argues that we should keep that option in mind.

Johnson described important details and background for each of these three arguments, and several others. Then she takes a step back and places the arguments in context by describing in detail fundamental ideas about technological development and about the concept of responsibility for that development. Johnson's argument focuses on the forward-looking nature of both technological development and responsibility. Looking ahead is prudent, but becoming fixed on a single possible future can lead us to ignore alternative (and perhaps better) possibilities. In particular, having a "vision" of future AI artifacts that are autonomous (relieving developers' responsibility for those artifacts) may lead to a self-fulfilling prophecy. An alternative to such a vision is a competing idea that all future AI applications should be developed in such a way that developers can always take responsibility for their creations. Johnson invokes STS principles that argue against any technological future as being inevitable, and instead insists that technologies are developed by human choices.

Johnson then concluded that "...whether or not there will ever be a responsibility gap depends on human choices not technical complexity." (2015, p. 714) Johnson points out that people can decide that the risks of some AI systems outweigh their possible benefits, and simply not use them (either as individuals or as political entities). The people who design, develop, and deploy these applications can be held accountable for the sociotechnical systems in which they function. This holding to account could act as a brake on the kinds of systems engineers will be willing to develop, since they could be at legal and ethical risk from any behaviors that society will not tolerate. However, Johnson makes it clear that any appropriate societal response to systems that seem to be out of direct human control will require deliberate, coordinated, and conscious effort to propose and implement.

Johnson Focuses on Algorithmic Accountability

We have seen some of the themes in Johnson's previous work that will reappear in her examination of algorithmic accountability. Next, we will focus on Johnson's 2021 paper, "Algorithmic accountability in the making," a recent publication on that topic (Johnson, 2021).

Algorithms Explained

First, we should clarify what an "algorithm" is. Johnson writes, "An algorithm is a sequence of computational steps that transform input into output." (Johnson, 2021, p. 113) I do not dispute that definition, but I will add a few illustrations and clarifications. An algorithm is different from a computer program that implements the algorithm. For example, there are many distinct algorithms for sorting a collection of data into ascending order. For simplicity, we will limit our attention to sorting a collection of numbers, but the same sorting algorithm can be used for sorting strings, or files, or anything else that has a defined ordering.

Perhaps the simplest sorting algorithm is known commonly as "the bubble sort." (The bubble sort happens to be notably inefficient for many cases of large collections, but that is not our concern here.) Figure 3.1 is a description of a bubble sort algorithm. (I use "a" and not "the" bubble sort algorithm, because there are many variations. Computer programmers would recognize this as a bubble sort, but they would also probably want to improve it with changes.)

The Fig. 3.1 description is expressed in "pseudocode," a notation that is similar to a computer program, but not identical. A computer program is written in a formal, well-defined programming language; pseudocode is similar but does not require the specificity of a true program. For example, in the pseudocode in Fig. 3.1, the word "swap" is not well defined. Programmers will know what "swap" means here, but most programming language compilers would not unless the programmer added more detail.

The pseudocode specifies (in a somewhat informal, but still clear expression) that the computer should execute a loop within a loop. The pseudocode assumes that the reader will know what an array is, how variables stand for values, and that individual values in the array can be accessed using indices. It assumes that the indices

Fig. 3.1 Bubble sort expressed in pseudocode

```
-------
Initialize N to the length of the Array to be sorted.
BubbleSort(Array, N):
    loop index i from 1 to N-1
      loop index j from 1 to N-i-1
        if Array[j] > Array[j+1]
          swap(Array[j], Array[j+1])
------
```

```
-------
def bubble_sort(Number_List):
    n = len(Number_List)
    for i in range(n-1):
        for j in range(n-i-1):
            if (Number_List[j] > Number_List[j+1]):
                Number_List[j], Number_List[j+1] = Number_List[j+1], Number_List[j]
    return(Number_List)
-------
```

Fig. 3.2 Bubble sort expressed in Python 3

Fig. 3.3 A bubble sort function in Java

```
----
static void bubbleSort(int[] arr) {
    int n = arr.length;
    int temp = 0;
    for(int i=0; i < n; i++){
        for(int j=1; j < (n-i); j++){
            if(arr[j-1] > arr[j]){
                temp = arr[j-1];
                arr[j-1] = arr[j];
                arr[j] = temp;
} } } }
----
```

go from 1 to N (not true in all programming languages). It also assumes that individual values in this array to be sorted can be compared for a characteristic "greater than," and that the reader will know what "swap" means when applied to members of the array.

Figure 3.2 expresses the algorithm in Fig. 3.1 as a Python 3 function. Unlike the pseudocode in Fig. 3.1, the Python 3 code in Figs. 3.2 *can* be transformed into machine language that executes on a machine.

Although the code in Fig. 3.2 faithfully implements the algorithm in Fig. 3.1, the Fig. 3.2 code includes details that are different. The loops use syntax that is specific to Python 3. In this Python 3 implementation, the numbers are kept in a list, not an array (the difference is chiefly how the numbers are stored "behind the scenes"). Python lists are indexed starting with 0, not 1. And the line…

```
swap(Array[j], Array[j+1])
```

…in Fig. 3.1's pseudocode is transformed into…

```
Number_List[j], Number_List[j+1] = Number_List[j+1],
Number_List[j]
```

… in Fig. 3.2's Python 3 code. Despite these (and other) differences, computer programmers would consider the code in Fig. 3.2 as a reasonably faithful implementation of the algorithm in Fig. 3.1.

Figure 3.3 expresses the Fig. 3.1 bubble sort algorithm in Java, a programming language different from Python.

Figure 3.3 is not identical to Figs. 3.1 or 3.2, but all three figures express the same algorithm. Figures 3.2 and 3 have details that are not made explicit in the algorithm of Fig. 3.1. For example, the Java code in Fig. 3.3 implements the "swap" of Fig. 3.1 using a temporary variable ("temp") that doesn't have an analog in either Figs. 3.1 or 3.2:

```
temp = arr[j-1];
arr[j-1] = arr[j];
arr[j] = temp;
```

To sum up, an algorithm is a plan for computation. Code implements that plan in a specific programming language with well-defined syntax, semantics, and meaning.

This distinction between abstract algorithms and concrete implementations has significant implications for accountability. When algorithms (which are at a higher level of abstraction) are implemented in a particular programming language (which are at a lower, more concrete, level of abstraction), details are necessarily added. There are responsibilities, and accountability, at both levels. There is a close relationship between an algorithm and code that implements it, but they are not identical. Assume that an algorithm A is devised in such a way that it favors males over females. Assume further that some code C is created to implement A on a particular machine. It is likely that C will replicate the male bias, but it is not inevitable. Furthermore, if A does not have such a bias, it is possible that a developer could add such a bias in C.

An algorithm and its implementation may have distinct ethical problems, and that possibility seems relevant to any discussion of developer responsibility. However, ethics literature has focused primarily on algorithmic accountability with respect to AI. We will follow that lead in this chapter, but with the understanding that implementations for AI algorithms will inevitably add another layer of complexity to all these issues.

Ways Algorithms Can Reflect Biases

Johnson lists different types of bias that Danks and London (2017) identified in algorithms: "training data bias, algorithmic focus bias; algorithmic processing bias; transfer context bias; and interpretation bias." (Johnson, 2021, p.112) Some biases can be blatant and easy to see in computer code. For example, imagine a program that examines data on job applicants, and screens out applicants based on demographic information. In a programming instruction such as this...

```
If (applicant[x].age > 59):
    Reject(applicant[x])
```

...the intent is explicit, and the ethical (and perhaps legal) issues related to age discrimination are obvious. This kind of straightforward bias surely exists in computer decision systems; for example, Volkswagen's diesel cheating scandal included code that explicitly cheated on emissions tests. (Davis, 2019). When algorithmic bias is obvious and serious, accountability is not complicated: whoever ordered, designed, coded, or approved that bit of computer code is responsible for it, and should expect consequences.

Another kind of bias can be introduced unintentionally by a programming error. Harking back to our discussion about algorithms and implementations, this could be based on an accidental introduction of bias as an algorithm is translated into a particular programming language. Here the responsibility most clearly not fulfilled is sufficient testing and quality control. Again, accountability seems clear: those in charge of the software are responsible for the problem. If and when it is discovered, at the very least, it should be their responsibility to correct the error, and perhaps provide recompense to those harmed by the unintentional bias.

Another source of algorithmic bias, based on data sets used to train AI applications, can be difficult to detect and correct. This difficulty is especially acute if AI applications are rushed to market to grab market share. Let us first take a mundane example, and then extrapolate to more value-laden examples.

Assume you are programming an AI application with the goal of recognizing chairs in images. Using a platform designed for machine learning, developers (perhaps hiring low wage workers for the task) present the AI application with thousands of images. In our hypothetical case, some of the images contain chairs, and some of which do not. For each image, a human tells the system whether the image contains a chair, and perhaps interactively draws a box around the chair if one exists in the image. This task is called "training" the system, and such a system and process are called "supervised learning," though the process has little resemblance to the training and learning that takes place with humans.

After many examples of images with and without chairs, the system can build up a complex system of internal memory that allows it to identify chairs inside of images. The inner workings of our "chair recognition system" can be difficult to interpret after training takes place. This is the problem of opacity that Johnson (among others) identifies as a potential problem for algorithmic accountability.

Back to our chair-recognizing AI system: imagine that no one thought to include pictures of bean bag chairs in the training set. It is unlikely that the system will successfully identify bean bag chairs as "chairs" in an image because of the data used to train the AI.

Although this weak chair-recognizing system might have some disadvantages, it seems unlikely that a major ethical problem will result. However, if a similar AI system was trained to recognize faces for use in disaster recovery, and if white, male faces were greatly overrepresented in the training set, then face recognition would be biased towards white males, and others might not get disaster help as quickly as the white males. Now the training set bias can have serious consequences with ethical significance.

Some AI systems are "trained" for users as the users interact with the systems. For example, a music playing service, or a video service keeps track of the music or videos we decide to watch. Using that data, the service can "guess" the kinds of media we prefer and offer suggestions for the next media to consume. Although we may or may not agree with these suggestions, if they are solely based on your previous choices, there is not an obvious unethical bias. However, if the service allows third parties to influence its suggestions, especially if this influence is not divulged to users, we can see important ethical issues. In this case, the algorithmic bias comes from the factors allowed to influence choices presented to a user. For example, if the suggestions are based in part on fees paid to the service (instead of on the users' preferences), a conflict of interest occurs.

Opacity, Transparency, and Why That Should Not be the Whole Story

We have described some systems here as "machine learning." A subset of machine learning systems are "deep learning" systems. (For a summary of the differences between other AI systems, machine learning systems, and deep learning systems, see (Kavlakoglu, 2020)). One way in which deep learning systems differ from other machine learning systems, is that deep learning systems typically require larger datasets for training. Another difference is that deep learning systems tend to be more complex in their inner workings. Both differences tend to make it more difficult for developers or users to explain (in any meaningful detail) how and why a deep learning system made a particular decision or choice. It may have been something about the data fed to the system; it may have been a detail of how that data was organized by the system; and it is probably both. Thus, AI systems (particularly deep learning systems) have attained a reputation for opacity, and that seems problematic for accountability.

The issue of explaining why an AI application generates a particular output (for example, a decision for a loan application, a potential medical diagnosis, or a recommended sentence for a convicted felon), is seen as central by many of those interested in algorithmic accountability. One way to describe this issue is to require that AI systems be "transparent," in contrast to the opacity described previously. However, transparency is in the eye of the beholder; a useful clarification is "transparent to whom?" For some computing systems, one explanation can make results transparent for the general public, and another (more detailed) explanation can make results transparent in a different way for a computer programmer. But for some AI systems, no explanation suffices for anyone, even AI experts.

Although she recounts how many people (both AI experts and others) are focusing on transparency to solve issues of accountability, Johnson insists that transparency is not the only aspect that must be addressed. That leads us to her detailed examination of accountability.

Accountability Explored

We have mentioned "accountability" previously, but it is useful to be more specific about that term now. In "Algorithmic Accountability in the Making," Johnson uses more than six pages explain "a broader account of accountability." (Johnson, 2021, pp. 117–123) I will not recite those details here, but instead will point out several themes. We have seen elements of each of these themes in Johnson's previous research.

Following Mark Bovens, Johnson defines accountability as "a relationship between an actor and a forum, in which the actor has an obligation to explain and to justify his or her conduct, the forum can pose questions and pass judgment, and the actor may face consequences." (Bovens, 2007, p. 450) But before applying this definition to algorithmic accountability, Johnson notes that there can be some problems.

First, Johnson points out that Boven's accountability definition is descriptive, not normative. If an actor does not agree that she has an obligation to (or a relationship with) a particular forum, does that mean no accountability exists? Because of this problem, and others, Johnson moves away from accountability as strictly a *relationship*, and towards a notion of accountability as a *practice* as well.

The emphasis on accountability practice is crucial to the phrase "in the making" in the title of (Johnson, 2021). Algorithmic accountability does not have standardized practices established, for software in general, and for AI software in particular. Johnson contrasts accountability in bridge-making, where social norms and accountability practices are firmly established, with accountability in algorithms, where norms and practices are not yet established. For algorithmic accountability, it is still being worked out who the actors and forums should be, and how practices of accountability (both formal and informal) should be established.

Johnson pushes beyond Boven's definition when she emphasizes the performative aspects of accountability. There must be performances by which actors and forums play out their roles in the relationship of accountability. Even when these performances are ritualistic (rather than heartfelt), the ritual still can reinforce the fundamental legitimacy of the assumptions of accountability. Performances are how the relationships between actors and forums are made concrete, and sometimes made public.

Relationships, norms, and performance are all important aspects of Johnson's description of accountability. Another aspect is the purpose of accountability; why, exactly, is accountability important? Johnson writes, "Accountability is generally understood to be directed at constraining power." (Johnson, 2021, p. 121) I contend that this is a particularly important emphasis in algorithmic accountability. The increasing influence of AI systems tends to centralize power (Brynjolfsson & Ng, 2023), and not having established accountability practices reduces society's effectiveness in properly constraining that power. Laws, regulations, and practices often lag behind technological novelty, and the fast pace of recent AI advances and adoptions are making this problem acute. (Greenstein, 2022; Boudens, 2007).

Accountability practices for algorithms are not keeping pace, especially for AI systems. Worrying about the slow pace of algorithmic accountability plays a part in the anxiety some AI researchers and practitioners are calling for caution, and perhaps delays in deploying AI systems. (Samuel, 2023).

Johnson's meticulous detail on the precise meaning of accountability is an example of "teasing apart" fundamentally important ideas in order to develop insights into complex questions about ethics and technology. I suspect that this kind of fine detail may be part of why many computing professionals find Johnson's work valuable. Computing thrives on careful, detailed work. Engineers in general, and software engineers in particular, are suspicious of overly broad generalizations, and significant assumptions hidden inside important conclusions. Johnson's care in defining terms finds an appreciative audience in many computing professionals.

Accountability Applied to Algorithms

Only after building a foundation of understanding about the fundamental nature of algorithms, and about important principles about accountability, are we ready to put those ideas together to better understand algorithmic accountability. Johnson makes the case that algorithmic accountability is only starting to be established. She lists four normative questions that need to be answered before we can claim that algorithmic accountability practices are no longer merely at an early stage of "in the making:"

1. Who should be identified and specified as the actors to be held accountable in algorithmic decision-making? What groups are appropriate forums for algorithmic accountability?
2. What shared beliefs and norms should be developed and promulgated?
3. Where and how should algorithmic accountability be performed?
4. What sanctions should be imposed on the actors in algorithmic decision-making?

We will examine each of these questions in turn.

Question 1, Part 1: Who Should the Actors Be in Algorithmic Accountability?

Johnson identifies a fundamental challenge for the first question: the number of people involved in algorithmic design, development, and deployment. This challenge, often characterized as the "problem of many hands," (Van de Poel et al., 2012) has well known applications in engineering ethics. This challenge to accountability can be met; indeed, many large engineering projects have "many hands," and have established norms and cultures that encourage and enforce accountability.

Johnson cites bridge construction as one example that includes many people, many kinds of expertise, and many possible causes of failure. Yet, when a bridge collapses, society expects and demands explanations in a public way. A recent example of this public expectation was seen in the U.S. when a major highway bridge collapsed in Philadelphia. (Wagh, 2023).

Simply because there is a public expectation of accountability does not guarantee timely fulfillment of that expectation. Take, for example, the long-playing controversies about the water supplied to residents of Flint, Michigan (Masten et al., 2016). It took years after activists asserted that the Flint water supply had become toxic for anyone to admit responsibility, and more years to determine the causes. Surely the many hands problem was part of the delay, as well as foot-dragging by many of the principal actors in the case. However, throughout that drama, it was never seriously questioned whether there should be accountability; it was assumed, because it is widely agreed that someone should be answerable for the quality of water delivered to residents.

To show how these two examples illustrate a problem with the current state of algorithmic accountability, imagine that the builders and maintainers of a bridge made the following claim: the design of this bridge allowed us to build it quickly, and to carry more traffic for less money than a traditional bridge; unfortunately, if the bridge fails, the design will make it difficult if not impossible to determine the exact cause of the failure. Again, consider a similar claim for those responsible for a city's water supply: we have made several decisions that reduce cost for delivering your water, and we were assured that water quality would not be affected; however, now that poor water quality has been detected, the design does not allow us to isolate the problem.

Of course, the public (its people and its institutions) would find those two claims about a bridge and drinking water completely unacceptable. If that bridge collapsed, or if water quality plunged, those claims would be dismissed out of hand, and rightly so. Committees would hold hearings, editorials would be published, lawsuits would be filed, people found responsible would be fired, and office holders would be voted out. Indeed, we have seen these kinds of consequences as people are held accountable for past disasters. This is the practice of accountability playing out for large, public engineering projects.

However, AI applications are being deployed (and accepted into widespread use) with the claim that transparency, explainability, and traceability are not practical (or even possible) with deep learning systems. We have plentiful examples of software systems that are not labeled AI that have failed. Examples include the crash of an Ariane 5 rocket (Dowson, 1997), a medical infusion pump that put patients at risk (Garde, 2014), and a social media "glitch" that erased some users' followers (Bennet, 2021). Thousands of examples are available in the Risks Digest, a decades-old forum about technology failures (Neumann, 2023). Some of those examples involved systems called AI. And authors have pointed out that AI systems are being designed in a way that makes them inherently risky (Hunt, 2023).

Which actors should be held accountable for algorithms and their implementation? At least a partial answer is the organization that profits from the system, and their employees who design, develop, and deploy those systems. Johnson lists companies, government agencies, and industries as other candidates for actors when those organizations are buyers, users, or participants in development. In particular systems, there may be other actors; for example, whoever provides training data becomes an actor with accountability. Despite the long list of potential actors, the fundamental question of who the human actors are seems clear. Accountability may have to be shared among players at different times and at different levels of abstraction, but that merely complicates the processes of accountability; it does not make them impossible.

Some people have suggested that when AI systems become sufficiently sophisticated, the AI systems themselves should be accountable for their behavior. If this suggestion merely asserts that harm due to a particular system should be identified and associated with the system, we agree. If the suggestion is to substitute accountability of the system to the people responsible for the system's deployment, Johnson strongly objects. Humans are the source of these systems. Humans decide to launch them. That act should require accountability, and the technical sophistication of the system does not somehow make that requirement disappear. Johnson wrote: "...although algorithms are essential components in algorithmic decision-making, they cannot be considered actors in accountability practices." (Johnson, 2021, p.124).

Another reason for rejecting algorithms as actors in the practice of accountability is linked to the ultimate purpose of accountability: constraining the exercise of power. If humans who launch AI applications can shift accountability to algorithms, then the humans' power remains unconstrained. The entire construct of accountability loses its potential for effectiveness.

Question 1, Part 2: What Groups Are Appropriate Forums for Algorithmic Accountability?

Johnson identifies the group of everyone affected by a system as a potential forum for algorithmic accountability. That group is rarely formally organized, but it can be effectively represented by government entities, the press, and activist groups organized around particular issues.

There are formal organizations, more specialized to the world of computing, that might become involved in the practice of algorithmic accountability. The Free Software Foundation, the Creative Commons, and the Wikimedia Foundation are examples of non-profit organizations that advocate regarding different aspects of computing. There are professional organizations such as the International Federation for Information Processing, the ACM, the IEEE Computer Society, and the Association for Women in Computing. There are organizations that focus on issues of technology and society, include IEEE's Society on Social Implications of Technology, the ACM's Special Interest Group on Computers and Society, and the International Society for Ethics and Information Technology.

All of these provide forums that could become involved in a process of accountability. Many of these organizations already play a role in professional ethics for computing professionals, and becoming involved in algorithmic accountability may be seen as a natural extension of their existing roles. This kind of involvement is already starting; for example, the ACM Technology Policy Council has published suggestions about generative AI (Leslie & Rossi, 2023). Staff and members of these groups could propose legislation and regulations, and act as expert witnesses to state and national governments. Using their members' professional contacts, the organizations could organize symposia and other meetings, out of which could emerge recommendations for either self-regulation or external regulations. In addition, the periodicals associated with these organizations could feature educational and editorial articles about algorithmic accountability.

Johnson points to the need for more formal, better articulated, and more widely accepted norms for algorithms. The organizations mentioned here as potential accountability forums could all act as sources or arbiters for such norms. However, none of the norms that might be used have, so far, gained broad acceptance. It is not the case that there is no activity in this area; in 2020 there were dozens of different AI-related codes of ethics proposed, (Hagendorff, 2020) and more have been published since. But none of those codes has gathered anything close to widespread acceptance. Part of the maturity to move from "in the making" to more well-established algorithmic accountability will require more consensus around fewer principles and documents. Existing organizations could become part of that movement towards consensus.

Question 2: What Shared Beliefs and Norms Should Be Developed and Promulgated?

Johnson notes the "striking" lack of norms for algorithmic design and use. (Johnson, 2021, p. 125) Johnson cites Mary Cummings (2021) as pointing out the sorts of issues that will have to be settled to develop useful norms in AI algorithms. Johnson makes a convincing case that such norms will have to be based on shared beliefs about the need for accountability for software, shared beliefs that have not yet been established.

Two examples of what may be emerging are the European Union's Artificial Intelligence Act, proposed in 2021 and adopted in 2024 (European Parliament, 2023); and the United States' Algorithmic Accountability Act, proposed in 2022. According to an analysis of the proposed laws by Mökander et al. (2022, p. 752),

"...they both aim to establish the governance infrastructure needed to hold bad actors accountable and allow actors with good intent to ensure and demonstrate that their ADS are ethical, legal and safe." As such, these legislative initiatives seem to indicate movement towards the kinds of norms that Johnson advocates.

Mökander et al. (2022) warn that the US legislation is unlikely to pass in the foreseeable future (due to the Democratic-Republican divide), and that the EU Act may not include sufficient detail to be immediately practical. Despite these warnings, the prominent discussions about algorithmic accountability offer at least some hope that Johnson's concern about the lack of norms for algorithmic design and use is starting to be addressed.

It is critical that a set of norms establishes limits on which kinds of applications and algorithms should be allowed in different contexts. This is a topic that the EU's AI Act addresses. It will be difficult to a priori regulate the design and development of software of any kind; the practicalities of such limits are daunting. However, the sale and deployment of software is more susceptible to regulation and enforcement, and harms that result from software systems after deployment can trigger accountability. At deployment, the context can be analyzed in a way that is difficult before deployment.

It is important that norms be established specifically for algorithmic accountability. But until that happens, problematic software can be sanctioned based on the software's use (or misuse) in systems that use software in a larger system that includes components that already have established norms for safety. There are fairly well-established existing norms for automobiles and airplanes, norms designed to improve the public good by encouraging safer systems. The norms do not prohibit designs or development, but they sometimes require particular processes and standards. For example, software used in commercial aircraft are subject to process requirements from several different agencies worldwide. (Sun et al., 2023). These norms are not always followed, but when unsafe systems result that cause harm, accountability is likely to follow.

The case of Boeing's 737 MAX airplane is an example that includes problems with software in a system that many other technologies as well (Konert, 2019; Johnston & Harris, 2019) Another example that is still playing out in courts, in government agencies, and in the public media is AI software that controls automobiles (Boudette, 2021). This "borrowing" of accountability norms from existing, mature industries helps establish responsibility for software failures. However, algorithmic accountability has unique challenges that will eventually require more specific, better-established norms focused solely on algorithms.

Algorithms are distinct from many other engineering artifacts. Algorithms are less concrete than the physical objects that civil engineers develop and the compounds that chemical engineers devise. Algorithms can be more complex than many physical devices (including the computers that the algorithms run on). This does not mean that meaningful accountability cannot be achieved for algorithms; it does mean that algorithmic accountability will require more than mimicking accountability in other engineering disciplines.

Question 3: Where and How Should Algorithmic Accountability Be Performed?

Making software safe and reliable is not a new problem. There is extensive literature on software engineering, quality control, testing, and software safety; techniques from this prior research could be applied to algorithmic accountability, and I think they should be. However, for many reasons, there has been resistance to requiring these techniques for AI applications. One argument is that AI techniques such as machine learning are different in kind from other kinds of software, and therefore these techniques are not appropriate. That argument seems reasonable, but inventing new variations of testing should not be impossible, only difficult.

Another reason given for not adopting more traditional quality control for AI is that it is likely that requiring these techniques will slow down development and deployment. The past pace of AI development, and the drive to secure large market share fast, fuels this impatience against more careful quality control. But many of us think that ethically this reasoning is weak, and that it does not serve the public well. This is why having accountability established is so important; if all players are held to appropriate standards of accountability, the motivation for avoiding accountability (and avoiding more stringent quality control) will be dramatically reduced.

Finally, many quality control procedures require an explicit, precise description of the desired outcome of software execution. For systems that change their behavior after deployment, effective quality control, both before and after deployment, becomes highly problematic. Our answer to this objection is that until AI developers get this problem under control, AI systems that change their behavior in unpredictable ways after deployment should be tightly controlled, if allowed at all.

One possible response to the problems of quality control for AI systems is to largely abandon quality control; we reject that response as irresponsible. Another response is to not deploy such systems unless quality control techniques are successfully adapted to them. These contrasting choices are a stark example of Johnson's contention that the course of algorithmic accountability is not inevitable; instead, the course will be determined by decisions at many levels.

Following the existing techniques for software quality control, processes for quality control should become standardized for algorithms throughout the software life cycle: during design, development, and maintenance after deployment. In the industry, techniques are known and sometimes practiced. However, there is not a universal acceptance of these processes, and they are not required, except in some specific areas such as software used in airplanes and in nuclear power plants. Making these requirements more broadly required would be a large step forward, although the processes for AI are still evolving. Both existing and future analyses should include both correctness (following a well-articulated specification of the desired behaviors), and software safety (Leveson, 1986).

When developers are convinced that a particular kind of software system resists current techniques for quality control, then new techniques should be developed before the system is released to the public. Some AI developers make claims that their software either is or will be resistant to current techniques of quality control;

ethical accountability requires that such software not be released before quality can be assured (both for the algorithms and training data), either by adapting existing techniques or developing new techniques (Zhang & Zhang, 2023).

The idea that AI systems should be delayed, perhaps for extended periods of time, may seem draconian. But an analogy suggests that it is sensible and prudent. Imagine that a new design for concrete and steel reinforcements is developed. Before products based on these designs could be used in construction, there would have to be many legal hurdles overcome, largely based on building codes in various jurisdictions. The delays to market for building materials is part of the cost of doing business. There are well-established norms and expectations that walls in new construction will not fall soon after they are built. If these norms and codes were ignored, and building subsequently collapsed, an outcry for accountability would likely be swift and sure. In fact, we see exactly this kind of scenario in the UK recently (Lee & Crew, 2023).

The idea that software systems are somehow exempt from such delays and costs seems, at best, special pleading from the people profiting from those systems. Surely as society becomes increasingly reliant on sociotechnical systems that incorporate AI software, then the safety of those sociotechnical systems becomes increasingly important. Society must recognize the importance of quality control throughout the software life cycle, and insist on it.

Question 4: What Sanctions Should Be Imposed on the Actors in Algorithmic Decision-Making?

Johnson (2021) does not recommend specific sanctions for algorithmic accountability. I will make some brief suggestions. Our answer to Question 2 indicates that norms have not been well-established for algorithmic accountability. However, there are already existing norms that have been enforced for sociotechnical systems that are not primarily software but do include software. We can, I think, look to these existing examples for suggested sanctions. When harm is determined to have occurred because of problems with algorithms, such sanctions could include monetary penalties, both to compensate people harmed and to discourage future malpractice. These penalties could be levied by regulatory agencies or by civil courts.

Public shaming is also a sanction that has occurred for lapses in technological systems. This shaming is carried out largely in the media, and competing narratives can muddle the effect. However, sometimes shaming has clear effects. Volkswagen and Boeing both were subjected to widespread criticisms for the "diesel-gate" scandal and the 737 MAX disasters, respectively. (Cohen, 2015; Prentice, 2022) The business consequences to both corporations were significant, and both corporations suffered significant monetary penalties.

In professions that require licensing, such as physicians and some engineers, sanctions for violating norms can include revoking the license to practice. Since computing professionals are typically not licensed, that sanction is not likely to be effective today. Should that situation change (and some continue to advocate for more widespread licensing of computer professionals (Isaak, 2022)), the revoking of licenses could be considered as sanctions for actors found to be negligent in the development of algorithms. The potential for such sanctions could be considered one argument for expanding licensing of computing professionals.

Perhaps private litigation for unsafe software will help establish norms. Perhaps legislation at various levels of government will begin to take effect. Again, attempts at laws, regulations, and lawsuits have been initiated (Congress.gov, 2022), but none of those actions has yet led to wide acceptance of well-recognized norms. As Johnson points out, the widespread acceptance of such norms will be necessary before algorithmic accountability can be well established.

Conclusion

Johnson's analysis of algorithmic accountability is useful in several ways, some of which are even more clear now than when she first published this work. First, Johnson's interest and analysis were years ahead of most philosophers, and almost all of the public. As a society, we were not nearly as concerned then as we are now. Johnson early recognition of these problems helped lay a foundation for scholarly work on these issues.

Since her earlier work was prescient, we should pay careful attention to Johnson's predictions that have not yet been borne out. Johnson's emphasis on sociotechnical systems (instead of focusing only on AI software) seems particularly important, as it runs counter to many contemporary approaches to these issues.

However thoughtful and thought-provoking they may be, Johnson's analyses are never intended as the *final* word. She invites us to go the next step forward, which often involves doing, not just identifying, the next right thing to do. Johnson's 2021 paper on algorithmic accountability is realistic, and sometimes even skeptical about the current state of the field; but the paper does not exhibit cynicism. Her emphasis on the human agency, following STS principles, is fundamentally optimistic, based on the belief that people can and should make better choices, choices that will benefit society.

For algorithmic accountability, those better choices involve advocating for norms that will assign appropriate responsibilities to appropriate actors. Those actors should include all the people and organizations involved in the design, development, and deployment of computing systems, including AI, machine learning, and deep learning systems. Those actors should *not* include artificial agents. The forums that will hold the actors to account should include non-profit organizations with expertise in software systems and professionals (such as professional organizations and associations), the press, governmental agencies, and legislatures, as well as the courts.

Together, the actors and forums should perform the rituals and processes of accountability. As those processes play out, the distinctive technical details of AI software should be analyzed, and the norms of algorithmic accountability are likely to be adapted to some of AI's idiosyncratic characteristics. But those distinctive characteristics should never be allowed to act as an excuse for abandoning accountability. That would be a tragic mistake, a mistake about which Johnson has given us fair warning.

Acknowledgments The author thanks the reviewers and editors for many helpful criticisms and suggestions. The author takes responsibility for any remaining shortcomings of this chapter.

References

Arkin, R. C. (2009). Ethical robots in warfare. *Technology and Society Magazine, IEEE, 28*(1), 30–33.

Bellah, R. N., Madsen, R., Sullivan, W. M., Swidler, A., & Tipton, S. M. (1996). *Habits of the heart: Individualism and commitment in American life*. Updated edition with a new introduction. University of California Press.

Bennet, C. (2021). *TikTok 0 followers glitch*. Dexerto. https://www.dexerto.com/entertainment/tiktok-0-followers-glitch-fix-for-account-bug-as-profiles-show-no-followers-1566312/

Borenstein, J., Grodzinsky, F. S., Howard, A., Miller, K. W., & Wolf, M. J. (2021). AI ethics: A long history and a recent burst of attention. *Computer, 54*(1), 96–102.

Boudette, N. E. (2021). Crashes involving Tesla Autopilot and other driver-assistance systems get new scrutiny. New York Times. https://www.nytimes.com/2021/06/29/business/tesla-autopilot-safety.html

Boudens, M. (2007). Analysing and assessing accountability: A conceptual framework. *European Law Journal, 13*(4), 447–468.

Bovens, M. (2007). Analysing and assessing accountability: A conceptual framework. *European Law Journal, 13*(4), 447–468.

Brynjolfsson, E., & Ng, A. (2023). Big AI can centralize decision-making and power, and that's a problem. In B. Prud'homme, C. Régis, G. Farnadi, V. Dreier, S. Rubel, & C. d'Oultremont (Eds.), *Missing links in AI governance* (pp. 65–87). UNESCO.

Chamberlain, J. (2023). The risk-based approach of the European Union's proposed artificial intelligence regulation: Some comments from a tort law perspective. *European Journal of Risk Regulation, 14*(1), 1–13.

Cohen, S. (2015). Volkswagen's shame and challenge to sustainability management. *Huffington Post*. https://www.huffpost.com/entry/volkswagens-shame-and-cha_b_8169830.

Congress.gov. (2022). H.R.6580 - Algorithmic Accountability Act of 2022. https://www.congress.gov/bill/117th-congress/house-bill/6580/text?q=%7B%22search%22%3A%5B%22Hr6580%22%2C%22Hr6580%22%5D%7D&r=1&s=1

Cummings, M. L. (2021). Rethinking the maturity of artificial intelligence in safety-critical settings. *AI Magazine, 42*, 6–15.

Danks, D., & London, A. J. (2017). Algorithmic bias in autonomous systems. In *Proceedings of the 26th international joint conference on artificial intelligence* (pp. 4691–4697).

Davis, M. (1991). Thinking like an engineer: The place of a code of ethics in the practice of a profession. *Philosophy & Public Affairs*, 150–167.

Davis, M. (2019). A whistle not blown: VW, diesels, and engineers. In A. Abbas (Ed.), *Next-generation ethics: Engineering a better society* (pp. 217–229). Cambridge University Press.

De Tocqueville, A. (2015). *Democracy in America – Vols. I and II*. Read Books Ltd.

Dowson, M. (1997). The Ariane 5 software failure. *ACM SIGSOFT Software Engineering Notes, 22*(2), 84.

European Parliament. (2023). EU AI Act: First regulation on artificial intelligence. *EU Parliament News*. https://www.europarl.europa.eu/news/en/headlines/society/20230601STO93804/eu-ai-act-first-regulation-on-artificial-intelligence

Garde, D. (2014). CareFusion runs into more 'deadly' trouble with infusion pump. *Fierce Biotech*. https://www.fiercebiotech.com/medical-devices/carefusion-runs-into-more-deadly-trouble-infusion-pump

Greenstein, S. (2022). Preserving the rule of law in the era of artificial intelligence (AI). *Artificial Intelligence and Law, 30*(3), 291–323.

Hacker, P., Engel, A., & Mauer, M. (2023). Regulating ChatGPT and other large generative AI models. In *Proceedings of the 2023 ACM conference on fairness, accountability, and transparency*.

Hagendorff, T. (2020). The ethics of AI ethics: An evaluation of guidelines. *Minds & Machines, 30*, 99–120. https://doi.org/10.1007/s11023-020-09517-8

Hellström, T. (2013). On the moral responsibility of military robots. *Ethics and Information Technology, 12*(2), 99–107.

Hunt, T. (2023). Here's why AI may be extremely dangerous—Whether it's conscious or not. *Scientific American*. https://www.scientificamerican.com/article/heres-why-ai-may-be-extremely-dangerous-whether-its-conscious-or-not/

Isaak, J. (2022). Get software right: license software engineers. *IEEE-USA Insight*. https://insight.ieeeusa.org/articles/op-ed-get-software-right-license-software-engineers/

Johnson, D. G. (2006). Computer systems: Moral entities but not moral agents. *Ethics and Information Technology, 8*, 195–204.

Johnson, D. G. (2007). Ethics and technology "in the making:" an essay on the challenge of nanoethics. *Nanoethics 1*, 21–30.

Johnson, D. G. (2011). Software agents, anticipatory ethics, and accountability. The growing gap between emerging technologies and legal-ethical oversight: *The pacing problem*, 61–76.

Johnson, D. G. (2015). Technology with no human responsibility? *Journal of Business Ethics, 127*(4), 707–715.

Johnson, D. G. (2021). Algorithmic accountability in the making. *Social Philosophy and Policy, 38*(2), 111–127.

Johnson, D. G., & John, M. M. (1995). Accountability and computer decision systems. *Communications of the ACM, 38*(12), 58–64.

Johnson, D. G., & Miller, K. W. (2008). Un-making artificial moral agents. *Ethics and Information Technology, 10*, 123–133.

Johnson, D. G., & Miller, K. W. (2011). Un-making artificial moral agents. *Ethics and Information Technology, 10*, 123–133.

Johnson, D. G., & Noorman, M. E. (2014). Responsibility practices in robotic warfare. *Military Review, 92*, 12–21.

Johnson, D. G., & Powers, T. (2005). Computer systems and responsibility: A normative look at technological complexity. *Ethics and Information Technology, 7*, 99–107.

Johnson, D. G., & Verdicchio, M. (2019). AI, agency and responsibility: The VW fraud case and beyond. *AI & Society, 34*, 639–647.

Johnston, P., & Harris, R. (2019). The Boeing 737 MAX saga: Lessons for software organizations. *Software Quality Professional, 21*(3), 4–12.

Kavlakoglu, E. (2020). AI vs. machine learning vs. deep learning vs. neural networks: What's the difference? *IBM Blog*. https://www.ibm.com/cloud/blog/ai-vs-machine-learning-vs-deep-learning-vs-neural-networks.

Konert, A. (2019). Aviation accidents involving Boeing 737 MAX: Legal consequences. *Ius Novum, 13*(3), 119–133.

Lee, D., & Crew, J. (Eds.). (2023). *Parents tell of shock as concrete fears force school closures*. BBC. https://www.bbc.com/news/live/uk-66461879

Leslie, D., & Rossi, F. (2023). Generative artificial intelligence. *ACM Technology Policy Council* (8). https://dl.acm.org/doi/pdf/10.1145/3626110

Leveson, N. G. (1986). Software safety: Why, what, and how. *ACM Computing Surveys (CSUR), 18*(2), 125–163.
Masten, S. J., Davies, S. H., & Mcelmurry, S. P. (2016). Flint water crisis: What happened and why? *Journal-American Water Works Association, 108*(12), 22–34.
Matthias, A. (2004). The responsibility gap: Ascribing responsibility for the actions of learning automata. *Ethics and Information Technology, 6*, 175–183.
Mökander, J., Juneja, P., Watson, D. S., & Floridi, L. (2022). The US Algorithmic Accountability Act of 2022 vs. The EU Artificial Intelligence Act: What can they learn from each other? *Minds and Machines, 32*(4), 751–758.
Neumann, P. (2023). RISK digest. https://catless.ncl.ac.uk/Risks/33/84
Prentice, R. (2022). *Engineering ethics and the Boeing scandal. Ethics unwrapped.* University of Texas. https://ethicsunwrapped.utexas.edu/engineering-ethics-and-the-boeing-scandal
Samuel, S. (2023). *The case for slowing down AI: Pumping the brakes on artificial intelligence could be the best thing we ever do for humanity*. Vox. https://www.vox.com/the-highlight/23621198/artificial-intelligence-chatgpt-openai-existential-risk-china-ai-safety-technology
Santoro, M., Marino, D., & Tamburrini, G. (2008). Learning robots interacting with humans: From epistemic risk to responsibility. *AI & Society, 22*(3), 301–314.
Sismondo, S. (2010). *An introduction to science and technology studies* (Vol. 1). Wiley-Blackwell.
Sun, J., Edwards, S., Connelly, N., Bridge, A., & Zhang, L. (2023). *Study on technical focuses and sampling coverage strategy of airborne software reviews*. https://www.academia.edu/download/81223906/13122ijsea02.pdf
Van de Poel, I., Nihlén Fahlquist, J., Doorn, N., Zwart, S., & Royakkers, L. (2012). The problem of many hands: Climate change as an example. *Science and Engineering Ethics, 18*(1), 49–67.
Wagh, M. (2023). The I-95 collapse is a tragic reminder that U.S. bridges are still built without fire safety in mind. *Popular Mechanics*. https://www.popularmechanics.com/technology/infrastructure/a44201682/what-could-have-prevented-i-95-bridge-collapse/
Zhang, J., & Zhang, Z. (2023). Ethics and governance of trustworthy medical artificial intelligence. *BMC Medical Informatics and Decision Making, 23*(1), 7.

Chapter 4
Johnson's Algorithmic Accountability and Corporate Accountability Dissonance

Kirsten Martin

Abstract In this chapter, I illustrate the impact of Johnson's scholarship on the study of corporate responsibility. I extend Johnson's accountability-as-practice to begin to scope (a) the normative grounding for *why* tech firms are accountable to their stakeholders and (relatedly) (b) *what* tech firms are accountable for, and (c) to *whom* firms are accountable. Firms are accountable for their design and deployment decisions about AI because firms have the power to make different decisions to elicit different moral implications in use. Firms are accountable for the decisions they make that impact others—whether those impacts are positive, as when creating value for stakeholders, or negative, as when firms destroy value for other stakeholders. Currently, industry has *accountability dissonance* where scholars and firms take credit for their ability to design algorithms that create value for key stakeholders while simultaneously shying away from the negative consequences, rules being broken, value being destroyed, or rights being diminished *for those same decisions.* I use online platforms to illustrate the importance of Johnson's approach to algorithmic accountability. In each case, firms have been slow to embrace accountability in the moral implications of their decisions; and attribution is made more complicated with the use of AI on a platform.

Keywords Accountability · Practices · Accountability dissonance · Online platforms

K. Martin (✉)
IT, Analytics, and Operations, University of Notre Dame, Notre Dame, IN, USA
e-mail: kmarti33@nd.edu

M. Noorman, M. Verdicchio (eds.), *Computer Ethics Across Disciplines*, Philosophy of Engineering and Technology 44,
https://doi.org/10.1007/978-3-031-87974-6_4

Introduction

Tech firms struggle to acknowledge the scope of their accountability in the technologies they develop and produce. For decades, the industry appeared to embrace Meta CEO Mark Zuckerberg's "Move fast and break things" mantra: negative consequences are a natural side effect for the best technologies (Hamilton, 2022). One cannot know, they claimed, how anything worked (Calvello, 2017; Knight, 2017). AI technologies perhaps were sentient or just super, super complicated (Huckins, 2023).

But then we kept hearing about social media platforms that targeted teens with mental health problems or that recommended an insurrection (Mac & Kang, 2021). The same technologies that were lauded as progress were suddenly seen as tools harming vulnerable communities. As an alternative, executives (with the help of a few philosophers) came up with a new theory known as longtermism or effective altruism (Ackermann, 2022; Torres, 2021). While firms may break things, and perhaps could do things differently, this new technology (whatever it is) is in long term interest of society—to include an indeterminate number of future sentient robots, future humans, etc. People may be harmed in the here-and-now, but AI technologies are going to make more people and robots happy in the future.[1]

Firms are dodging and weaving to avoid being seen as accountable for the technologies they produce. The shapeshifting narrative morphs to fit whatever counter argument or experience is reported: technology is neutral,[2] accurate,[3] efficient, and we all benefit from its existence either now or in the long, long term. Firms either had no say in the matter (technological determinism) or their brilliance led the design decisions to create the best technology for everyone (Martin, 2022a). Or, firms make mistakes but the magic of the market 'picks' the best technology. The ebb and flow of this industry narrative has only one constant theme: firms are not accountable for any downsides of their technology.

Enter Deborah Johnson's article on Algorithmic Accountability (2021). Johnson's accountability in the making reframes accountability as a relationship between an actor (e.g., firm) and a forum of others who believe an actor has an obligation to account to them: "accountability is a social practice involving actors, forums, shared beliefs and norms, performativity, and sanctions, and the practice is aimed at putting constraints on the exercise of power." Importantly, whether firms acknowledge their role and accountability is not particularly important to holding firms accountable. In fact, firms' denials may just be a part of a process or practice of accountability.

Johnson's algorithmic accountability allows us to see industry's shape shifting narrative about their (lack of) accountability as part of an accountability process: a

[1] Unfortunately, those future humans and robots will then be sacrificed for even-more-future humans and robots if longtermism is applied at that time.

[2] Algorithms, like all technology, are not neutral (Johnson, 2023; Martin, 2022a)

[3] Firms may create that accuracy for profit (Martin, 2023a) However, industry still claims it is accurate and efficient (MIT Technology Review Insights, 2023).

forum demands actors to be accountable for their actions, actors may reflexively deny their role or accountability (or the harms), yet the accountability discourse should continue (see auto safety, manufacturing pollution, opioid epidemic, tobacco industry, etc.).

In this chapter, I focus on the important implications of Johnson's approach to technology and accountability for research in management and business ethics. Where the AI accountability discourse can fixate on adequate notification to subjects or users, Johnson rightly (and consistently (Johnson, 2011, 2015; Johnson & Miller, 2006; Johnson & Powers, 2005)) refocuses accountability discussions on the value-laden decisions of developers of technology as creating technology with moral implications. As such, technology, including AI, are moral entities within their larger socio-technical system, and developers are accountable for their decisions in the creation of those moral entities—including when their decisions have moral implications for stakeholders of the AI in use (Johnson, 2011). Johnson, in this manner, leads a group of scholars arguing that technology is value-laden through design, and firms are accountable to a forum for these value-laden decisions (Abebe et al., 2020; Akrich, 1992; Birhane et al., 2021; Cummings, 2006; Johnson, 2023; Latour, 1992; Martin, 2019; Rosenblat & Stark, 2016).

The goal of this chapter is to explore the implications of Johnson's accountability in the making for firms, markets, and industries. I illustrate the impact of Johnson's scholarship on the study of corporate responsibility and extend her accountability-as-practice to begin to scope (a) the normative grounding for *why* tech firms are accountable to their stakeholders and (relatedly) (b) *what* tech firms are accountable for, and (c) to *whom* firms are accountable. These are related—identifying *why* firms are accountable to their stakeholders will shine a light on not only *what* firms are accountable for but also *to whom* they are accountable. Firms make value-laden decisions in the technologies they bring to market or technology they adopt. And firms are accountable to their stakeholders—those impacted by the firm's decisions or impacting the firm—as to the moral implications of their decisions (Martin, 2019). Specifically, I argue:

- Accountability is a check on those in power (Johnson, 2021). Firms are accountable for their design and deployment decisions about AI because firms have the power to make different design/deployment decisions to elicit different moral implications in use. Firms are accountable to those impacted by those decisions because (a) firms have power to choose their decisions (b) which have consequences for others.
- Johnson's approach to algorithmic accountability broadens *to whom* firms are accountable by emphasizing that firms are accountable to those that are impacted by decisions made in the design, development, and implementation of algorithms. Vulnerable stakeholders to the firm—legitimate stakeholders impacted by a firm's decisions about AI but without power in the market or with the firm—feel the sharp edge of AI since the tool is disproportionately used to predict and categorize less powerful individuals in society (Benjamin, 2019; Eubanks, 2018; Martin, 2022b).

- Similarly, Johnson's approach bring clarity as to *what* decisions firms are accountable for. Firms are accountable for the decisions they make that impact others—whether those impacts are positive, as when creating value for stakeholders or negative, as when firms destroy value for other stakeholders. Currently, firms have a form of *accountability dissonance* where scholars and firms take credit for their ability to design algorithms that create value and a positive impact on key stakeholders while simultaneously shying away from the negative consequences, rules being broken, value being destroyed, or rights being diminished *for those same decisions.*

I use online platforms to illustrate the importance of Johnson's approach to algorithmic accountability in piercing through a *fog of accountability*.[4] In each case, firms have been slow to embrace accountability in the moral implications of their decisions; and attribution is made more complicated with the use of AI on a platform. These platforms allow other actors to transact, and those actors sometimes act in ways that are unethical. However, as I show below, Johnson's approach to algorithmic accountability sheds light on the specific accountability of platforms—to whom they are accountable, why they are accountable, and for what actions are they accountable—even with the creation of what we may call a fog of accountability created not only by the use of AI but also by the involvement of other market actors contributing to the problem.

Johnson's Accountability in the Making

In "Algorithmic Accountability in the Making," Deborah Johnson takes a sharp turn away from algorithmic accountability as an issue of opacity (2021). Johnson argues this current focus on opacity and transparency "reflects a narrow and inadequate understanding of accountability" (p. 113). Similar to Nicholas Diakopolos, transparency may promote accountability, but transparency does not ensure accountability.[5] While a lack of opacity may be an issue for AI (as it is for many organizational decisions), Johnson argues that solving the opacity problem through 'transparency' or 'explainability' does not solve the issue of accountability: *who is accountable for the AI being used and to whom?*

Johnson begins with Mark Bovens' broader account of accountability as a starting point where Bovens describes accountability as "a relationship between an actor and a forum, in which the actor has an obligation to explain and to justify his or her

---

[4]Where Johnson and Powers note the added complexity with computer systems, with hardware and software and the many organizations that design and develop different parts of the technology, here I focus on the many human actors on a platform and the difficulty in parsing out who is accountable for what. According to actor-network theory, we would making similar arguments about different actors (material and nonmaterial) (Johnson & Powers, 2005).

[5]Transparency is "an enabling factor that can support the monitoring of behavior with respect to ethical expectations" (Diakopoulos, 2020, p. 197).

conduct, the forum can pose questions and pass judgement, and the actor may face consequences" (Bovens, 2007).

Johnson identifies the important element of this definition of accountability as being within a relationship between an actor and a 'forum' where a forum is "any group of individuals who believe an actor has an obligation to them" (p. 118). To understand accountability, according to this approach, begs the question *accountable to whom?* For business, which will be important for AI, Johnson notes, "A company, for example, may be accountable (for different types of behavior) to a board of directors, regulatory agencies, employees, and customers" (p. 188).

Within the field of management, strategy, business ethics, etc., we refer to these members of the forum as *stakeholders* (Freeman, 1984, 2010; Freeman et al., 2007, 2020). While the field has shifted since, traditionally stakeholders are defined broadly as those who are impacted by a firm's decisions or impact the firm (Freeman, 1984). And, firms are accountable to stakeholders: "managers must account for the interests of these stakeholders when making decisions" (Phillips et al., 2003, p. 485).

Freeman, the seminal stakeholder scholar within the field, uses the terms responsibility or obligation in a manner similar to Bovens' and Johnson's use of accountability: both phrases are used to describe how an actor (a firm) must answer to others (stakeholders) and, for Freeman, how decisions the firm makes should take into consideration those stakeholders, since the firm will be accountable to them. Key to all is that the accountability emanates from the relationship between the actor and the forum (Bovens, 2007; Johnson, 2021) or between the firm and stakeholders (Freeman, 2010).

For Bovens, and differing from Johnson, both actor and forum *share* the belief that the actor is accountable to others.[6] Johnson rightly points out (and business ethicists would agree!) that "there are situations in which actor and forums disagree. This situation typically occurs when an actor does not believe that an explanation is owed while a forum does" (2021, p. 118). Johnson notes this is where she and Bovens diverge:

> On Bovens's account it would seem that we would have to say that there is no accountability because the actor and forum do not share the belief that an explanation is owed. This, however, doesn't seem right. That is, it seems wrong to say that when an actor doesn't believe he or she owes an explanation, the actor is, therefore, not accountable (p. 118).

Johnson identifies the key implication for where we are with AI: "it is important to note that when actors and forums disagree, it does not mean necessarily that there is no accountability, rather it may be a sign of accountability practices *in the making*." This approach to accountability should resonate with business scholars. Firms are accountable to many stakeholders—actors in a forum for Bovens and Johnson—with whom they have a relationship (Freeman, 1984, 2010). Disagreement as to the scope of what firms are accountable for does not mean firms have no accountability but only that accountability practices are *in the making*.

[6] "A central aspect of the relationship between actor and forum is that both parties share the belief that the actor has an obligation to explain particular types of behavior" (Johnson, 2021).

We can therefore take as a starting point to understand firms' accountability to stakeholders of their decisions about AI:

> accountability is a social practice involving actors, forums, shared beliefs and norms, performativity, and sanctions, and the practice is aimed at putting constraints on the exercise of power (Johnson, 2021, p. 123).

Johnson's approach, however, will broaden to whom firms are accountable (and for what) since Johnson's approach does not require the actor in the forum (or stakeholder) to have an established relationship with the firm. Key stakeholders for algorithmic accountability are those "who are or might be affected by algorithmic decision-making."[7]

Broadening the Scope and Grounding of Business Ethics

Why Firms Are Accountable: AI and Power

Power—who has it, how it is deployed and to what effect—is critical to understand AI. First, as noted by Jonne Maas, AI evokes a "power-dependence relation between those who shape a system (i.e., developers and users) and those affected by a system (i.e., end-users)" (Maas, 2023). Second, firms operate in markets where not all actors have equal power (Poole et al., 2021). Firms are accountable because their decisions, made possible through their power in the market and within this power-dependent relation between developers and users, impacts others. For Johnson, accountability is a practice "aimed at putting constraints on the exercise of power."[8]

We should expect to see accountability-as-practice in situations where power is not currently being held to account or where firms are acting in a manner that negatively impacts less powerful stakeholders. These marginalized stakeholders—by definition—do not have the power to hold the firm accountable. In fact, this lack of power is why the firm is able to make opportunistic decisions, where the harms fall on marginalized stakeholders.

For those that study firms and markets, we can take from Johnson:

- Firms should reasonably expect to be held accountable for their actions and decisions by those who are or might be affected by algorithmic decision-making.
- Firms will be held accountable *in particular* in those situations where they exert power over stakeholders through the design and deployment of algorithms.
- Ignoring or denying accountability for actions is one step in being held to account—such denials are not the final word but an expected part of the process.

---

[7] "When it comes to the forums for algorithmic accountability, perhaps the largest forum is all of those who are or might be affected by algorithmic decision-making" (Johnson, 2021 p. 125).

[8] See also Maas, accountability is meant to place a check on domination or superior and unaccountable power (Maas, 2023).

- We will see accountability in the making in situations where firms abuse power to harm vulnerable stakeholders since more powerful stakeholders are able to push back on firms when they are harmed and the 'negotiation' is done in private.

This last point is important to understand how *little* reliance we should place on firms' denial of accountability for their actions. Since accountability is a tool to place restraints on the abuse of power, calls for accountability will occur if and only if firms' exercise of power is not being adequately restrained by the immediate stakeholders. When we see accountability in the making, we should assume that a more powerful actor is currently not being held to account for the harms inflicted on less powerful members of the forum. Their denial should be expected.

Clarifying What Decisions Firms Are Accountable for

For business ethics and management generally, Johnson's approach to algorithmic accountability has implications for the scope of what firms will be accountable for. In other words, in correctly identifying firms as being accountable for their decisions about AI because (a) these firms are in power to make decisions that (b) impact others, we can then think through the scope of what firms should be held accountable for. Specifically, firms have the power to design, develop, and use AI. And firms are able to make different design and deployment decisions with different moral implications in use. Firms decide the training data to use, how data is labeled, the assumptions and models to use, the outcome variable to optimize, the types of testing before and after deployment, the criteria for 'success' in testing, etc., etc.

Any decisions that impact stakeholders—including value-laden design decisions of AI or the decisions to adopt and deploy AI within an organization—are within the scope of what firms are accountable for. This is not different from stakeholder theory that says that firms are accountable for the decisions they make, for the value they create and destroy, for the rights they enable or diminish, for the rules they break or respect, and for the values they uphold (Freeman et al., 2020).

And, firms take credit for design decisions or adoption decisions for the value these decisions create for themselves and some stakeholders. Their design decisions create value for the firm and are touted as a competitive advantage. People should use their product, work at their firm, and invest in their vision due to their strategic decisions to design and develop and adopt the 'right' AI technology. The executives and managers take credit for their decisions around AI and are happy to be held accountable for the positive impacts of their decisions.

Simultaneously, firms deny accountability for the negative implications for these same decisions. Firms have claimed a type of *algorithmic accountability dissonance*: firms make value-laden decisions in regards to the design, development, or adoption of AI and want to be accountable for the good outcomes, but claim they are not accountable for negative outcomes *for the same decision* (Fig. 4.1).

Firms waive away the moral implications of the same design decisions for which they wish to take credit. For example, cheating detection software exclaims the

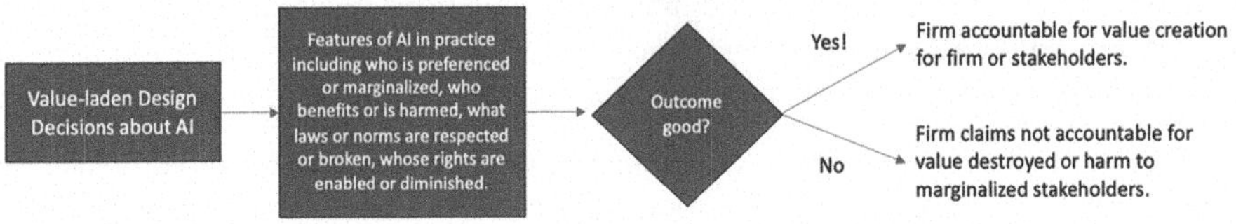

Fig. 4.1 Algorithmic accountability dissonance

value their design creates for schools and educators while brushing past the negative impacts to falsely accused students. Resume readers boast their ability to find the right candidate while blaming others for the same technology discriminating against women.

In general, if a firm claims to create value from their decision and wishes to be held accountable for the creation of value from their actions, then the firm is also accountable for the moral implications of those decisions as well. The accountability dissonance emerges when a firm wishes to be considered the 'decider' or 'actor' held accountable for the *good* consequences or moral implications of their decision while, perhaps, shrugging their shoulders as to the negative moral implications of that *same* decision.

Broadening To Whom Firms are Accountable

So far, I have argued that Johnson's approach to algorithmic accountability augments the stakeholder approach to corporate responsibility. Similar to stakeholder theory, firms are accountable to stakeholders for the decisions over which they have power (that are in their control). And more focus and more accountability will be thrust upon those firms with more power since, according to Johnson, accountability is a practice meant to constrain the exercise of power. Johnson applies this approach specifically to algorithms. Firms are accountable for the value laden decisions they make in regards to technology in the same way firms are accountable to stakeholders for their decisions around the design of cars and auto safety or are accountable to stakeholders for their decisions around how they treat employees.

Within management, the stakeholder approach to corporate responsibility, in its current incarnation, tends to focus on those stakeholders with whom firms have an immediate relationship or contract (customers, suppliers, communities, employees, financiers, etc.) which is a subset of stakeholders firms are accountable to for their actions. For Johnson, a key stakeholder—or member of the forum—to whom firms are accountable are those impacted by AI decisions, such as subjects and users of AI. However, these stakeholders do not have a direct or market relationship to the firm developing or deploying AI.

In fact, the original approach to stakeholder theory within business ethics and strategy defined stakeholders broadly as *those impacted by or impacting firm decisions* (Freeman, 1984). However, more recent scholarship has focused on those stakeholders who are voluntary, with whom the firm has a relationship or agreement, or who previously provided value to the firm (Freeman & Phillips, 2002; Phillips, 1997; Van Buren III, 1999). This more limited definition focuses on those stakeholders that are obvious to managers.

This is a problem (Dawkins, 2015; Greenwood & Mir, 2019; Martin, 2023b). By limiting the scope of firm accountability to those stakeholders with whom the firm already has a relationship, who are voluntary, and who—in some way—have provided value to the firm, scholars studying firms are missing the less powerful, vulnerable stakeholders who have legitimate claims and interests. These vulnerable stakeholders—such as data subjects, victims of hate speech online, users or targets of AI programs—are too easily ignored in the current version of management theory. Where management theory tells Facebook to pay attention to regulators and advertisers, users as well as the subjects of hate speech, targets of online trolls, and victims of a recommendation algorithm pushing non-consensual porn are left stranded with their legitimate claims but no market power.

Johnson's approach should remind stakeholder theorists that firms are accountable to actors and organizations that are impacted by their decisions *regardless* if there is a contractual relationship or if the stakeholder is in a voluntary transaction with the firm. In fact, for AI, subjects and users of AI—e.g., students subject to cheating detection programs, individuals scanned at the border using FR—have no direct relationship with the firm developing AI but are considered key actors in the forum to whom the firm is accountable.

Within management theory and practice, this narrow definition of stakeholder, limiting obligations to those with whom a firm has a relationship and who previously benefited the firm, may be descriptive in what firms regularly do but lacks a focus on what firms *should* do in terms of legitimate stakeholders who are directly impacted by an AI decision but who have little power. In fact, those legitimate, marginalized stakeholders, impacted by a firms' algorithmic decisions, are the people most in need of a firm being held accountable to them given the power-dependent relation between those who shape a technology and the end users who are subjects of the technology (Maas, 2023). In other words, Johnson's approach pushes stakeholder theory to return to its normative roots and ask *to whom should a firm be accountable* and answer *those who are impacted by the firms' decisions.*

Platforms and the Fog of Accountability

Johnson's approach to accountability in the making provides clarity as to not only why firms are accountable but also for what and to whom firms are accountable. Firms are accountable because they use their power to make AI design and development decisions impacting others within the forum. Firms are accountable for decisions over which they have power (and could have decided otherwise) and to stakeholders, broadly defined, impacted by those decisions.

I turn now to demonstrate the utility of this approach to algorithmic accountability through digital platforms, where the actions of the platforms are sometimes minimized in comparison to the actions of the actors on the platform. Platforms act as an exchange and create a market for other organizations, individuals, and market actors to transact (Martin, 2024). The NYSE, Tinder, eBay, Facebook, and Google Search are all platforms or exchanges. Platforms have become more and more prevalent in our lives and, at the same time, have proven slippery when we look for who is accountable for what happens on a platform. Tech firms, reporters, and scholars appear to frame platforms as 'neutral' or as even having the goal of neutrality—in that platforms only facilitate the actions of others on their platforms.[9] For platforms, the difficulty in attributing accountability in the design and deployment of AI is exacerbated by the difficulty in identifying the actions attributable to the platform versus the actors on the platform.[10]

For example, normally prices of goods or services sold on the platform are not the responsibility of the platform, since actors on the platform set their own prices for the goods and services they sell. However, Amazon recently designed a program to set and move prices on their own platform. Amazon's use of a program to set prices on a platform illustrates the fog of accountability introduced by the use of an algorithm to facilitate a decision and the vagueness of the role of a platform in facilitating the transactions of other market actors. The normal ambiguity around accountability is made more complicated—creating a fog of accountability—if the firm can hide behind (a) the algorithm and (b) the mistaken notion that platforms are 'neutral' facilitators.

Social media platforms, for example, point to users and content creators as the bad actors who should be held accountable for the content created—e.g., dehumanizing, anti-LGBTQ hate speech. However, social media companies are accountable for their decisions about not only what content to keep on their site but also what content to recommend. Social media platforms are accountable for those decisions where they have power and to individuals impacted by those decisions. So, both can be true: the user is accountable for creating the hate speech the social media platform is accountable for promoting and recommending the hate speech by those impacted by the promotion—whether current users on the platform or not.

Travel Reservation Platform

Take, for example, a travel reservation platform. The platform allows travel agents (in the old days) or consumers (recently) to find and reserve flights from many different airlines. Such a travel platform does not dictate the routes that airlines offer or the types of seats or even the price. The platform may require that all prices be in

[9] "The myth that online platforms are neutral pervades the tech industry" (Chander & Krishnamurthy, 2018).

[10] Similar to the complexity due to hardware versus software as noted by Johnson and Powers, and the problem of many hands as noted by Villegas-Galaviz, here I am arguing that platforms introduce an accountability fog due to many different actors (material and non-material) with not-clear lines of accountability (Johnson & Powers, 2005; Villegas-Galaviz & Martin, 2023).

USD or that the type of information provided by airlines is consistent to make sure that consumers can easily compare. These types of design decisions help facilitate the exchange or market for flights and decrease the cost to transact for consumers and airlines by being able to find and compare flights. So, we can think of platforms as making some design decisions and taking some actions for which they should be accountable.

Table 4.1 shows how we may start to think about the actions, decisions, and behaviors of different actors associated with digital platforms—for which they are accountable. In each case, airlines and travel exchanges claim to create value from these actions—a travel exchange will claim to show you the best routes for your preferences or make it easy for you to find a new travel route—but also are held accountable for the negative impacts or moral implications of those same decisions. Sabre, for example, was held accountable for their recommendation of American Airlines above all other airlines even when American was not the best route or price for the consumer (Friedman & Nissenbaum, 1996; Martin et al., 2024). Travel platforms consider their recommendation system to be a competitive advantage or

Table 4.1 Platforms, algorithms, and accountability

	Actors on the Platform		Platform Decisions
Travel Platforms	Airline Decisions: pricing, routes, services	Customer Actions: payment, rating,	Travel Platorm: recommendation, formatting, vetting, etc. what user data necessary to collect, use, share
Dating Apps	User Actions: feedback, truthfulness in profile, choosing potential partner, communication with other users.		Dating Platform: Factors of users to include, prioritize, exclude; recommendation system, ability to block/protect, vetting, etc.; what user data necessary to collect, use, share.
Social Networks	User Actions: Whether to engage with content online, what content to post.		Social Networks: What posts and groups to recommend and prioritize for users; which posts to remove; what engagement options to make available (like, dislike, post, etc); what reminders to provide; what feedback to allow (reporting of content); what user data necessary to collect, use, share
Search Engine	User Actions: Prompts, search terms, choice of link for content, engagement with content provider	Content Providers: proving relevant content for users.	Search Platform: what content to include in the search index; prioritizing search results; matching users to relevant content; deprioritizing or blocking content could be fraudulent; what user data necessary to collect, use, share
LLM (ChapGPT)	User Actions: Prompts to provide, how to use the answers, whether to attribute the LLM in the use of the answer provided, etc	Content Providers: Whether to allow their work to be used in model development.	LLM: which sites or sources to use for training, design decisions as to model assumptions, which sites or sources to *exclude* from model development, which prompts to exclude from model development, how to apply the model in use, *what answers to provide for prompts*.

differentiator for users, and the same platform must also be held accountable for the negative implications of those same recommendations.

Dating Apps

Dating apps, such as Tinder, eHarmony, or Bumble, offer a platform for individuals to find romantic matches. The actors on the platform are seen as the primary decision makers—deciding who they wish to engage with, what information is provided, how each individual is 'portrayed' on the app, and whether or not they wish to continue to communicate or see someone they meet.

However, the dating platform makes design decisions as to the type of information that is required or even possible to share. In doing so, these platforms decide which factors are important to consider when looking for a particular mate: whether height or education is important; whether marital status is required, whether truthfulness is even a goal. These platforms also decide which individuals on the site are in control of who they hear from or who users can block or mute. In addition, these dating apps decide the order in which potential partners are recommended to the users, which factors to highlight as particularly important, which factors to prioritize in the recommendation system. Dating platforms differentiate and compete based on these types of decisions.

Importantly, decisions as to what information is allowed to be important in choosing a potential romantic partner, in addition to how potential partners are sorted and prioritized, are value-laden decisions for which dating apps are accountable to their users. For example, dating apps use of race as a factor has come under criticism as prioritizing intra-racial dating and even segregating people by race or ethnicity—which has a long, problematic history in the United States (Bedi, 2019, 2022; Williams, 2024). These same value-laden recommendation and sorting decisions are posited as a competitive advantage and differentiator—dating apps claim to be *better* at facilitating a match—and it holds that these same dating apps are accountable for the moral implications of those same decisions.

Generative NLP Models (v. Search)

Bert and Chat GPT offer another example of how firms may mistakenly not acknowledge accountability for their design decisions—particularly when used to augment search. These tools create new content using generative language processing models. So, where a search engine allows users to enter prompts and then provides related links to relevant content (websites, government documents, etc), Chat GPT *creates* new content to answer the same type of prompt provided by the user while also providing links to original content. The new content is generated by Chat GPT, and Chat GPT was created by 'learning' different types of content deemed acceptable by Open AI (owners of Chat GPT). Open AI had to decide which type of content was appropriate to include in training data and input to Chat GPT's models. And Open AI decided the use cases for their technology as well as how 'answers' are presented to users. Also, Open AI decided whether source material was also provided to users or not.

While the use of LLM is search is positioned as mere facilitator, similar to a search platforms, LLM design decisions are above and beyond the decisions that a search engine would have to make. As such, Open AI, through the design and use of Chat GPT, makes more value-laden decisions than a search platform and even claims to have a competitive advantage based on those previous and current decisions. OpenAI should also be accountable for these same value-laden decisions.

Conclusion

This chapter has sought to illustrate how Johnson's algorithmic accountability as practice sheds light on why, to whom, and for what firms are accountable when it comes to AI. Firms are accountable to stakeholders because their decisions have moral implications for stakeholders. Design and development decisions around AI, digital platforms, facial recognition, etc. create value for stakeholders including the firm. These decisions are why users flock to their technology, the rationale for charging users and customers for their technology, and the argument for why people should invest in their company: because these firms are making hard, important decisions in the design and development of their technologies and those decisions lead to positive outcomes. Firms are not reluctant to being accountable for the good outcomes of design and development decisions and make sure that those decisions are attributable to their unique capabilities. However, these decisions also undermine rights, marginalize vulnerable stakeholders, foster abuse, have negative consequences, break norms, rules, and laws, and create unfair systems and outcomes. Within this practice of algorithmic accountability, more work needs to be done to (a) not take seriously tech firms claims they are not accountable for the moral implications of their decisions and (b) help those same firms work through the value-laden design decisions as well as identify the moral implications of those decisions.

References

Abebe, R., Barocas, S., Kleinberg, J., Levy, K., Raghavan, M., & Robinson, D. G. (2020). Roles for computing in social change. In *Proceedings of the 2020 conference on fairness, accountability, and transparency* (pp. 252–260).

Ackermann, R. (2022, October 17). The growing influence of effective altruism. *MIT Technology Review*. https://www.technologyreview.com/2022/10/17/1060967/effective-altruism-growth/

Akrich, M. (1992). The de-scription of technological objects. In W. Bijker & J. Law (Eds.), *Shaping technology/building society: Studies in sociotechnical change* (pp. 205–224). MIT Press.

Bedi, S. (2019). Online dating sites as public accommodations: Facilitating racial discrimination. In S. Brison & K. Gelber (Eds.), *Free speech in the digital age*. Oxford University Press.

Bedi, S. (2022). Sexual racism. In *The Palgrave handbook of sexual ethics* (pp. 407–419). Springer.

Benjamin, R. (2019). *Race after technology: Abolitionist tools for the new Jim code*. John Wiley & Sons.

Birhane, A., Kalluri, P., Card, D., Agnew, W., Dotan, R., & Bao, M. (2021). The values encoded in machine learning research. In *2022 ACM conference on fairness, accountability, and transparency* (pp. 173–184) https://arxiv.org/abs/2106.15590

Bovens, M. (2007). Analysing and assessing accountability: A conceptual framework 1. *European Law Journal, 13*(4), 447–468.

Calvello, A. (2017). We will never fully understand how AI works—But that shouldn't stop you from using it. *Institutional Investor*. https://www.institutionalinvestor.com/article/2bstr1aauoiex25yto45c/opinion/we-will-never-fully-understand-how-ai-works-but-that-shouldnt-stop-you-from-using-it

Chander, A., & Krishnamurthy, V. (2018). The myth of platform neutrality. *Georgetown Law Technology Review, 2*, 400.

Cummings, M. L. (2006). Integrating ethics in design through the value-sensitive design approach. *Science and Engineering Ethics, 12*(4), 701–715.

Dawkins, C. (2015). Agonistic pluralism and stakeholder engagement. *Business Ethics Quarterly*, 1–28.

Diakopoulos, N. (2020). Accountability, transparency, and algorithms. The Oxford handbook of ethics of AI, 17(4), 197.

Eubanks, V. (2018). *Automating inequality: How high-tech tools profile, police, and punish the poor*. St. Martin's Press.

Freeman, R. E. (1984). *Stakeholder management: Framework and philosophy*. Pitman.

Freeman, R. E. (2010). *Strategic management: A stakeholder approach*. Cambridge University Press.

Freeman, R. E., Harrison, J. S., & Wicks, A. C. (2007). *Managing for stakeholders: Survival, reputation, and success*. Yale University Press.

Freeman, R. E., Parmar, B. L., & Martin, K. (2020). *The power of and: Responsible business without trade-offs*. Columbia University Press.

Freeman, R. E., & Phillips, R. A. (2002). Stakeholder theory: A libertarian defense. *Business Ethics Quarterly*, 331–349.

Friedman, B., & Nissenbaum, H. (1996). Bias in computer systems. *ACM Transactions on Information Systems (TOIS), 14*(3), 330–347.

Greenwood, M., & Mir, R. (2019). Critical management studies and stakeholder theory: Possibilities for a critical stakeholder theory. In *The Cambridge handbook of stakeholder theory* (pp. 35–52). Cambridge University Press.

Hamilton, I. A. (2022, February 16). Mark Zuckerberg's new values for Meta show he still hasn't truly let go of "move fast and break things". *Business Insider*. https://www.businessinsider.com/meta-mark-zuckerberg-new-values-move-fast-and-break-things-2022-2

Huckins, G. (2023, October 16). Minds of machines: The great AI consciousness conundrum: Philosophers, cognitive scientists, and engineers are grappling with what it would take for AI to become conscious. *MIT Technology Review*. https://www.technologyreview.com/2023/10/16/1081149/ai-consciousness-conundrum/

Johnson, D. G. (2011). Computer systems moral entities but not moral agents. *Machine Ethics, 168*.

Johnson, D. G. (2015). Technology with no human responsibility? *Journal of Business Ethics, 127*(4), 707–715. https://doi.org/10.1007/s10551-014-2180-1

Johnson, D. G. (2021). Algorithmic accountability in the making. *Social Philosophy and Policy, 38*(2), 111–127.

Johnson, G. (2023). Are algorithms value-free? Feminist theoretical virtues in machine learning. *Journal Moral Philosophy, 21*(1–2): 1-35. https://philpapers.org/rec/JOHAAV

Johnson, D. G., & Miller, K. W. (2006). *A dialogue on responsibility, moral agency, and IT systems* (pp. 272–276).

Johnson, D. G., & Powers, T. M. (2005). Computer systems and responsibility: A normative look at technological complexity. *Ethics and Information Technology, 7*(2), 99–107.

Knight, W. (2017, April 11). The dark secret at the heart of AI: No one really knows how the most advanced algorithms do what they do. That could be a problem. *MIT Technology Review*. https://www.technologyreview.com/2017/04/11/5113/the-dark-secret-at-the-heart-of-ai/

Latour, B. (1992). Where are the missing masses? The sociology of a few mundane artifacts. In W. Bijker & J. Law (Eds.), *Shaping technology/building society: Studies in sociotechnical change* (pp. 225–258). MIT Press.

Maas, J. (2023). Machine learning and power relations. *AI & Society, 38*(4), 1493–1500.

Mac, R., & Kang, C. (2021, October 3). Whistle-blower says Facebook 'chooses profits over safety'. *The New York Times*. https://www.nytimes.com/2021/10/03/technology/whistle-blower-facebook-frances-haugen.html

Martin, K. (2019). Ethical implications and accountability of algorithms. *Journal of Business Ethics, 160*(4), 835–850.

Martin, K. (2022a). Algorithmic bias and corporate responsibility: How companies hide behind the false veil of the technological imperative. In K. Martin (Ed.), *Ethics of data and analytics*. Taylor & Francis.

Martin, K. (2022b). *Ethics of data and analytics*. Taylor & Francis.

Martin, K. (2023a). Predatory predictions and the ethics of predictive analytics. *Journal of the Association for Information Science and Technology, 74*(5), 531–545.

Martin, K. (2023b). Who counts in business ethics. *Business Ethics Quarterly, 33*(1), 216–243.

Martin, K. (2024). Platforms, privacy, and the honeypot problem. *Harvard Journal of Law & Technology, 37*, 3.

Martin, K., Hong, G., & Easley, R. (2024). When platforms act opportunistically: The ethics of platform governance. *Working Paper*. https://papers.ssrn.com/sol3/papers.cfm?abstract_id=4202821

MIT Technology Review Insights. (2023, October 5). "Transformative" AI will lead to rapid 25% efficiency gain, according to 81% of tech execs surveyed for new MIT Technology Review Insights research report. *PR Newswire*. https://www.prnewswire.com/news-releases/transformative-ai-will-lead-to-rapid-25-efficiency-gain-according-to-81-of-tech-execs-surveyed-for-new-mit-technology-review-insights-research-report-301948000.html

Phillips, R. A. (1997). Stakeholder theory and a principle of fairness. *Business Ethics Quarterly, 7*(1), 51–66.

Phillips, R., Freeman, R. E., & Wicks, A. C. (2003). What stakeholder theory is not. *Business Ethics Quarterly, 13*, 479–502.

Poole, S. M., Grier, S. A., Thomas, K. D., Sobande, F., Ekpo, A. E., Torres, L. T., Addington, L. A., Weekes-Laidlow, M., & Henderson, G. R. (2021). Operationalizing critical race theory in the marketplace. *Journal of Public Policy & Marketing, 40*(2), 126–142.

Rosenblat, A., & Stark, L. (2016). Algorithmic labor and information asymmetries: A case study of Uber's drivers. *International Journal of Communication, 10*, 27.

Torres, E. (2021, October 21). Against longtermism. *Aeon*. https://aeon.co/essays/why-longtermism-is-the-worlds-most-dangerous-secular-credo

Van Buren, H. J., III. (1999). If fairness is the problem, is consent the solution? Integrating ISCT and stakeholder theory. *Academy of Management Proceedings, 1999*(1), C1–C6.

Villegas-Galaviz, C., & Martin, K. (2023). Moral distance, AI, and the ethics of care. *AI & Society, 39*, 1–12.

Williams, A. (2024). Not my type: Automating sexual racism in online dating. In *Not my type*. Stanford University Press.

Chapter 5
Opening the Black Box of AI, Only to Be Disappointed

Jonne Maas and Jeroen van den Hoven

Abstract The common approach to AI Ethics is concerned with developing and deploying non-harmful AI systems by aiming to align the systems with commonly acknowledged ethical principles and values. However, this approach has been criticized for neglecting the broader societal context in which these systems and values are embedded. We argue that due to this neglect, the standard approach to AI Ethics remains normatively disappointing. For one, a narrow construal of AI Ethics risks some form of political naivete. Two, by leaving a larger normative context out of consideration it fails to show why we have moral reasons to act on the output of an AI system, everything considered. A normatively satisfying account of AI Ethics requires what Langdon Winner calls a 'political ergonomics' of AI in which larger politico-economic questions regarding society, politics, and power distribution are part and parcel of the lifecycle of an AI system. Integration of normative political theory and AI Ethics is therefore essential.

Keywords AI ethics · Politics of AI · Deontic powers · Langdon Winner · Political ergonomics

J. Maas (✉)
Delft University of Technology, Delft, The Netherlands

The Green Land, Utrecht, The Netherlands

J. van den Hoven
Delft University of Technology, Delft, The Netherlands
e-mail: M.J.vandenHoven@tudelft.nl

M. Noorman, M. Verdicchio (eds.), *Computer Ethics Across Disciplines*, Philosophy of Engineering and Technology 44,
https://doi.org/10.1007/978-3-031-87974-6_5

Introduction

In his paper "Upon Opening the Black Box, Only to Find it Empty", Langdon Winner (1993) critiqued social constructivism. Social constructivism is a theoretical framework used in Science and Technology Studies (STS) to describe how a technological artefact has come into being in a social setting. This framework has provided a wealth of insights of how different stakeholders, perspectives, and interests are involved in the development and deployment of a particular artefact. It shows how artefacts have been shaped in and by social interactions, and at which inflection points in the history of the origin of technologies different turns in their development could have been taken.

While an important and valuable framework, social constructivism has its limitations. As Winner points out, the framework provides a historical and social science point of view that only *describes* the development of a particular technology, thereby neglecting the normative implications and assumptions underlying technological innovations. Winner argues that, while it is surely helpful to open the black box of technology and understand how technological artefacts have come about, if we neglect the moral dimension of their development, important political and power aspects in their genealogy remain hidden.

Winner's observation regarding social constructivism contains an interesting lesson for our thinking about AI Ethics. AI Ethics aims to align AI systems with commonly acknowledged ethical principles and values, with Explainable AI (XAI) being one of the holy grails of the AI era. Ideally, XAI will realize other values such as accountability and safety as it allows for a better understanding of the system. Although AI Ethics clearly addresses moral dimensions and is therefore not as normatively empty as Winner considers social constructivism to be, numerous scholars have voiced concerns about the current state of AI Ethics (Birhane et al., 2022; Crawford, 2021; Gebru, 2022; Mittelstadt, 2019; Selbst et al. 2019; Van Maanen, 2022). In particular, much critique centres on AI Ethics' neglect of the broader societal and political context.

A vivid example of the worries of a narrow AI Ethics is portrayed by Keyes et al. (2019, p. 3). They discuss the disturbing algorithmic software that is supposed to manage overpopulation and food shortages by turning lonesome elderly into nutritious food such as "hash browns (Grandmash™)" and "bananas (Nanas™)," a process the authors refer to as "mulching". The authors apply the FAccT (Fairness, Accountable, and Transparent) framework in order to see how we can develop and deploy this system more responsibly. This satire points out how a narrow construal of AI Ethics can sometimes fail to address highly relevant questions. While most initiatives in AI Ethics may perhaps not leave us completely empty-handed, they often do leave AI Ethics in a normatively disappointing state.

This disappointing state can be characterized in the following two ways. First, in its current form AI Ethics is often politically naïve. We distinguish between a simple version and a complex version of political naivete. Simple naivete quite straightforwardly neglects the broader societal perspective, such as when we praise volunteers and soup kitchens for the poor, without addressing rampant inequality in society. Complex naivete can take the form of limitations of conceptualizations of moral values. Common mistakes include a failure to take into account a moral dimension of a concept, such as when we work with a statistical notion of fairness, e.g., fairness as equalised odds, without taking alternative conceptions of fairness into consideration.

Another example is to ignore the wealth of fundamentally different conceptions of the value one is considering. Thinking about digital democracy requires consideration of democratic conceptions which each foreground different normative key concepts such as representation, information, contestation, deliberation, or participation. These different conceptions each point to different technological 'democratic' innovations and technological development trajectories. Simple and complex political naivete then risk developing ineffective methodologies and tools to achieve ethical AI.

Second, a narrow construal of AI Ethics is disappointing as its main focus is on 'what' the system does, i.e., whether it is aligned with ethical values and principles, whereas it neglects 'how' the system has been brought into society. Drawing on work by John Searle (1995), this 'how' question can be seen to be important as systems have a particular deontic status and are embedded in a normative context and in a network of deontic relations. The AI Ethics literature is less concerned with what provides these systems their status, and therefore what normative grounds we have ultimately to base our actions on the output of AI systems. In order to fully appreciate the moral reasons we may have for doing so, we need a clear picture of what we call the *deontic provenance*—i.e., the normative pedigree of AI's status in society—of the relevant AI we are confronted with. Mapping the deontic provenance of AI brings in view the normative dimension of how an AI system is embedded in a larger politico-legal context, such as the rule of law, human rights, and democracy.

To improve on both points, the naivete and the lack of a clear picture of the deontic provenance of AI, we need better alignment between society and technological innovations, or what Winner (1987) calls "political ergonomics" in the analysis and design of Technology. In Sects. 2 and 3 we discuss in more detail our two concerns with a narrow construal of AI. In Sect. 4 we propose that a political ergonomics along the lines suggested by Winner is much needed to open the black box of AI in a normatively satisfying manner, and in Sect. 5 we conclude.

Political Naivete in AI Ethics

The AI Ethics literature discusses moral implications of AI systems. It has spawned numerous ethical guidelines that have been developed to highlight important values and principles that systems should meet (e.g., transparency, autonomy-preserving, privacy-safeguarding, see Jobin et al., 2019; Hagendorff, 2020; Floridi & Cowls, 2022). These values and principles now feed into proposals for regulatory frameworks like the AI Act currently under development by the European Commission (2021).[1] It also includes the immense collection of studies concerned with technical and practical attempts to design AI systems better aligned with values and principles that are often presented at AI Ethics conferences such as FAccT and AIES. Think of such more technical and practical attempts proposals to increase explainability (Gunning et al., 2019) and fairness (Hirota et al., 2022). These initiatives broadly aim to generate Responsible AI systems from a principlist standpoint, which according to the European Union's plans for AI governance and regulation imply technical robustness, compliance with extant law, and important moral values and human rights.[2]

However, as several scholars have remarked, the dominant approach to AI Ethics rooted in principlist ideals often remains narrow, shallow, and compartmentalized.[3] Deborah Johnson (2021) points out that striving for XAI by simply suggesting technological solutions does not capture the political, economic, and social complexity of a digital society. This results in the inability to achieve and realize values like accountability in practice. Others have argued that the literature fails to come to terms with the fact that a normatively satisfying approach to ethical AI requires institutional and structural change (Gebru, 2022), and must position the development and deployment of AI—including the accompanying values—in its relevant societal context. Without a contextual awareness, we run the risk of falling into what Selbst et al. (2019) refer to as 'abstraction traps' (e.g., inaccurately applying a system designed for one context to a different one, or neglecting how new technologies affect and change societal behaviour and values).

Birhane et al. (2022) have empirically investigated these worries and claims. The authors annotated a total of 526 papers part of important AI Ethics conferences (FAccT and AIES) from 2018 to 2021 along the three following lines. First, they identified their abstract/concrete engagement with ethics (including concepts like bias, fairness, etc.) applied to a specific context. Second, they assessed to what extent a FAccT or AIES paper mentions the disparate impact of AI. And third, in case of strong, medium, or weak disparate impact mentions, the authors assessed to what extent these disparate impacts were made explicit. Although more towards 2021 the authors noticed that the AI Ethics research increasingly connects to broader issues, their overall findings support worries and claims that the dominant AI Ethics literature as presented at FAccT and AIES remains rather abstract and conceptually

[1] The link to the EU's plan to regulation shows how a particular conception of doing AI Ethics (i.e., a principlist one) feeds into and shapes regulatory measures.

[2] Of course there are numerous approaches to AI Ethics, and some more aligned with what we believe a normatively satisfying account of AI Ethics would look like. Our criticism is specifically directed towards more principlist approaches to AI Ethics. In Sect. 4, we further discuss different conceptions of AI Ethics.

[3] Such principlist ideals are characterized by the need to conform to certain principles such as explainability, fairness, or accountability, and are represented well by e.g., Floridi and Cowls (2022). The satire by Keyes et al. (2019) mentioned in the introduction highlights neatly the discomfort reasoning from isolated principles.

ambiguous on ethical values and principles, and indeed insufficiently acknowledges societal context.

The current approach to AI Ethics, then, "[shrinks] ethics into one-dimensional problems," focusing strongly on ambiguous values and principles that get translated into technological design solutions (Gerdes, 2022, p. 2). In the mulching satire mentioned in the introduction, for instance, the authors propose to increase accountability by giving potential elderly-to-be-mulched a ten second window to object the algorithm's decision (Keyes et al., 2019). 'Accountability' is then reduced to a ten second window, and does not include the political, economic, and social complexity that Johnson emphasizes.

The example of installing this short window shows how a narrow and compartmentalized construal of AI Ethics is problematic as a one-dimensional AI Ethics results in a shoppable 'checklist ethics' (Bolte et al., 2022) that can easily be repurposed and cherry-picked by more powerful agents in society. One issue with this is that it leaves AI Ethics in its current state prone to 'ethics washing' (Van Maanen, 2022). However, it furthermore risks being politically naïve, which we can understand in at least two ways.

First, we can be naïve simply by overlooking the broader societal context in which the AI system is developed and deployed. For example, the whole point of developing a solution for the filter bubble and content moderation issue on social media platforms would be undermined by the system being deployed in a context in which the state practices internet censorship. Similarly, we can wonder what 'democratic design' meaningfully contributes if a system will be deployed in a dictatorship. Consider efforts in sustainability. Clearly, an airline company will not make much headway on sustainability by asking passengers to bring their own re-usable coffee mugs. Such 'micro-focuses' steer towards a solutionist ideal—one where technology can fix all issues (Morozov, 2014), if only they are designed ethically. However, these are examples of non-solutions for much larger and more complex political concerns. Studies of professional codes of conduct and micro-ethics issues will thus not get us far if they do not address macro questions as well.

Second, there are also more complex versions of being politically naïve related to the ambiguity of principles and values. For instance, we may focus on preferences as economists do, but forget to ask how our preferences are manipulated, falsified, and adapted. These then fall short of serving as reliable indicators of what people have reason to value. This would for example ignore Steven Lukes' (2021) third dimension of power—i.e., the power to shape preferences, desires, and will formation—that he distinguishes from his first and second dimensions of power to affect decisions and the power to set agenda's, respectively. Without explicit attention to this third dimension—which seems particularly pertinent to critique the role of social media in Big Nudging and computational propaganda and cognitive warfare—one could responsibly and carefully take stakeholder's views into account without considering how these views have been shaped by the dominant socio-technological infrastructures.

Another form of complex naivete occurs when conflicting conceptions of one concept are insufficiently considered. Maas (2022), for instance, has argued that AI

Ethics remains too narrowly focused on Berlin's negative conception of freedom that is aligned with Mill's harm principle (i.e., you should be free to act as long as you do not cause harm to anyone). However, when we take on a different perspective on 'harm', one can be morally wronged even when one is not interfered with actively. Such moral wrong—known as 'domination' in contemporary political philosophy of a neo-republican type—relates to a different conception of freedom than the more traditional freedom as non-interference commonly associated with liberal democracies. A change in what precisely constitutes harm matters for how we go about designing systems. We must hence be weary of complex naivete, given that how we conceptualize specific values and principles influences how we design and develop technology (Veluwenkamp & van den Hoven, 2023).

Both forms of political naivete draw our attention away from the fact that law, ethics, and AI are all already situated in a particular complex socio-political, economical, and institutional setting that may undermine or support mechanisms to protect and further democratic ideals. Indeed, technology is situated in a particular societal context, being shaped by and shaping society (Bijker, 1995; Johnson, 1997; Winner, 1980). Given this interplay between technology and society, democratic values are at stake when we fail to reflect critically on the complexity and dynamics of an (increasingly globalized) society. Consequently, such failure may potentially result in an inability to translate ethical insights into meaningful pragmatic methodologies and tools (Carnegie Council, 2021; Gebru, 2022).

Worries about compartmentalization in applied ethics have already been voiced in other areas such as sustainability development and climate justice. Lehtonen (2004, p. 212), for instance, argues that we need a "coevolutionary framework" to resolve difficult trade-offs in sustainability development, and Simon Caney advocates for the "method of integration" over the "method of isolation" (Brandstedt, 2014, p. 73). As Caney argues, whereas isolationism detaches the debate around climate change from other normative considerations such as poverty, the method of integration advocates a holistic approach that includes other concerns besides climate justice. Climate change can, for instance, result in an increase in climate refugees, so we may need to re-evaluate our migration rights (Brandstedt, 2014, p. 74). Both Lehtonen and Caney recognize that different spheres of life inevitably influence each other and contend that viewing conflicting normative concerns in isolation therefore makes a methodological error that will never truly lead to normatively satisfying solutions.

AI Ethics currently misses such a co-evolutionary framework and integrationist approach. As is increasingly shown by scholars in the field, AI Ethics initiatives including the guidelines, regulatory proposals, and the FAccT and AIES literature have remained on a level where they do not provide us the bird's-eye perspective necessary to address the larger political, social, economic, and legal concerns. And so, we can be normatively disappointed in the current state of the art in AI Ethics due to both its simple and complex political naivete.

The Missing Debate on *Deontic Provenance*

There is a second reason why the neglect of the broader societal picture gives us reasons to be normatively disappointed. AI systems are inevitably intertwined with our social lives and institutions (Gabriel, 2022). They complement or replace human agents, producing a particular output we act upon. In other words, drawing on Searle's (1995) theory on the construction of social reality, we are constructing the social world with AI. Whereas AI Ethics is seriously concerned with 'what' makes AI systems ethical, there is less of a focus on 'how' the system has come to be part of our social world. However, this 'how' is fundamental. In Searle's terms, constructing our social world necessarily grants these systems an assigned and accepted 'deontic status', which, as we claim, co-determines the moral and practical reasons we may have to act on its authority. As will become clear in this section, in order to normatively assess our reasons to act in the company of AI systems, we thus need to know the *deontic provenance* of the relevant digital entity.

This requires representing the entity's normative pedigree, which cannot only be done via opening the black box of the system by means of internalist perspicuous representations such as XAI provides. It is also a matter of understanding of how the black box was delivered on our doorstep, where the system got its status and authority from, which is an externalist perspicuous representation. Before we look at the quality of the content of the box, let's inspect the bill of lading. We do not mean to suggest that the proposed broadening of the scope replaces the work that is done in the field of XAI and should be abandoned. Yet to anticipate the next section, we just suggest that work on an internalist representation needs to be supplemented along the lines suggested by Winner's political ergonomics, which meaningfully situates a technology in a normative analysis with a broader scope.

Searle (1995) points out that certain entities (objects, individuals, or groups of individuals) in society have a particular moral standing because the function of their status comes with certain normative or 'deontic powers', i.e., they have certain rights, duties, and obligations that are directly linked to their particular status function. Think, for instance, of the exclusive right the President of the United States has to order the use of nuclear weapons. Or, a scrap of paper with a distinctive ink pattern printed on it that counts as money. A status function with deontic powers, or 'deontic status', is imposed on the piece of paper, and from there on the paper figures in a network of deontic relations of rights and permissions. There is a collective acceptance, recognition, and acknowledgement of the deontic powers of relevant actors and institutions, e.g., the national bank, to confer such status functions and to turn scraps of paper into legal tender.

In his work on status functions and deontic powers, Searle leaves aside the complex societal background that supports the conferral of this status or status function imposition. However, this provides a limited perspective on the *normative* construction of our social world, in particular our socio-technical world, as a status function with deontic powers has some form of authority that gives us reasons to act on it. After all, when we act on a twenty-dollar bill, we act on the fact that it counts as *money*, not on the fact that it has written 'twenty dollars' on it. Without its status function, we would have no reason to give back change, for instance. The conferral of deontic status is thus relevant to normatively assess our relation to the status function.

To make this more concrete, consider a railway timetable. We can study a piece of paper with rows and columns of train numbers, departure and arrival times and stations, etc. In and of itself such a piece of paper is practically useless and lacking deontic authority. However, if the timetable was issued by the proper institutional party (e.g., British Rail) with the official institutional powers associated with it, the timetable now suddenly gets its deontic authority from the fact that it is the *official* timetable.

Clearly, there is a difference between 'just' a piece of paper with columns and rows vs. the 'official' timetable, even if they would have the exact same information on it. If we would find both papers on the ground, we would base our actions on the official timetable because it is more evidently linked to other status functions, such as the British Rail. Indeed, *because* it is the official timetable, we have moral and practical reasons to act on it. Practical, because it is a more reliable indicator of the British Rail's timetable. Moral, because if we were to share information with others, we ought to share the information presented by the *official* timetable. There are thus certain entities of which we need to know their *deontic provenance*—i.e., the pedigree of the entity's deontic status—in order for us to determine their authority and reliability.[4]

We can understand deontic provenance along the lines of the inheritance principle. Formally put, the inheritance principle implies that "if A inherits φ then there is some Δ that is a source of A's φ-ness (i.e. A inherits φ from Δ and no entity among Δ inherits φ)" (Trogdon, 2018, p. 5). Taking money as an example for φ, Schaffer (2016, p. 95, footnote and emphasis omitted) writes that:

> (…) a grounded entity inherits its reality from its grounds, and where there is an inheritance there must be a source. One cannot be rich merely by having a limitless supply of debtors, each borrowing from the one before. There must actually be a source of money somewhere.

In the context of Searle's work on the construction of the social world, this means that there must be money that is accepted as legal tender, there must be a contract and normative expectations that borrowers pay back their loans, etc. Indeed, money does not come from nowhere, it requires a particular normative context and is embedded in a complex network of deontic relations. Deontic provenance is then to make explicit how status function A inherited its deontic status φ from which source of 'higher' deontic status Δ.

Just as in a piece of paper with rows and columns there is nothing intrinsically authoritative, there is nothing in the output of an AI system—or in the technology producing it for that matter—*by itself* that can explain why we would have moral and practical reasons to act upon it. To consult it multiple times, or to consult a

[4] Of course, there are other reasons to act on an entity's inherent properties. A comb cannot be used as a chair, for instance, whereas a tree stump may.

slightly modified version of it will also not help. It will have to get its deontic status from elsewhere in order for us to have moral reasons to act on its output. Indeed, it needs a source of higher deontic status.[5] It is one thing to look at an AI system in the legal or health care system and ascertain that it has certain desirable properties of transparency and fairness, but we also need to know whether it has been approved to play a particular role in decision making in these contexts. Without such a status function imposition no amount of assurance concerning its moral properties justifies its utilization.

We find this additional normative dimension has received insufficient attention in the AI Ethics literature. Of course, some work points in this direction. For instance, scholars like Binns (2018) and Dobbe et al. (2021) have voiced their worries about the social power of AI developers,[6] and have suggested ways to ground this power by building on political ideals like public reason and deliberation. Approaches that aim to ground developers' power clearly hint that something is 'off' with development and deployment of AI. By more clearly delineating the deontic provenance of AI systems, we may well better understand precisely *what* is off. This should then also shed light on whether the approaches such as those suggested by Binns and Dobbe et al. in fact resolve the issue.

In addition to other regulatory measures such as the GDPR, the AI Act more directly aims to sketch out the deontic provenance of an AI system. The Act asks AI developers to provide extensive record keeping and documentation, so that we can "assess the compliance of the AI system with the relevant requirements" (European Commission, 2021, p. 30). The idea, here, is that if the system complies with the relevant requirements mandated by the AI Act, we can trace back its deontic provenance via compliance with the Act to the European Union, an established political institution. In other words, documentation is essential to show that the system is backed up by a politico-legal institution, where an AI system inherited its deontic status from the AI Act, which in turn inherited its status from the European Union.[7]

[5] Some clarifications are in order. First, we refer in particular to 'high-sensitive' AI systems, such as those deployed in legal, medical, financial, and public governance sectors. Second, as suggested in footnote 2, there are other reasons to act on a system's output. A high accuracy, for instance, could give us moral and practical reasons to act on a system's output. Accuracy, however, focuses on how the system itself behaves, yet not on how it was developed and deployed. A system's deontic status, then, gives us a distinct set of moral and practical reasons to act. Third, where a deontic status *gives* us certain reasons, deontic provenance allows us to *assess* whether we have these reasons. We thank Herman Veluwenkamp for discussion on these points.

[6] With social power of AI developers, we refer to the fact that developers make certain design choices that influence the system's behaviour, which in turn has an effect on the system's end-users. There is therefore an indirect dependency relation between the system's end-user on the developers (via the AI system) in the sense that end-users to some extent depend on the developers with respect to how the system will behave. We consider this dependency relation a social power relation. See Maas (2023).

[7] This raises at least three further questions. It requires us to ask, first, where we begin with the source of deontic power; second, whether deontic provenance can be traced back to a single source; and third, whether 'deontic provenance' is simply conflated with 'lawfulness'. We believe deontic provenance need not have one source. For example, there are also national-level regulatory mea-

In theory, the AI Act can thus be an effective mechanism to outline the deontic provenance of AI systems. In practice, however, we see the AI Act is criticized for being too heavily influenced by industry (Galvagna, 2023; Perarnaud, 2023; Perrigo, 2023; Schyns, 2023) and for being negligent in oversight regarding the fact that AI developers can decide which risks are acceptable (Burri, 2022; Laux et al., 2023). These criticisms point to a further direction regarding deontic provenance, namely whether the AI Act in fact provides a suitable normative source for AI systems from which to inherit their deontic status.

Indeed, some important yet underdeveloped questions in the literature in this line of research are (1) what does it mean for AI systems to have a normatively substantiated and well-founded deontic provenance, and (2) are those requirements currently met? Although these questions are beyond the scope of this chapter, we are encouraged that taking into consideration the broader societal context can in fact shed light on these issues. For now, we merely wish to point out that AI Ethics is too strongly fixated on 'what' makes a system ethical, thereby neglecting the 'how' the system's deontic status was assigned, accepted, and preserved.

Towards a Normatively Satisfying Account of AI Ethics

As we have seen, we have at least two reasons to be normatively disappointed in AI Ethics as typically practiced today for its neglect of the broader societal context in which AI systems are necessarily situated. First, it risks being politically naive as it may overlook important questions at the heart of democratic societies, which ultimately may lead to developing ineffective methodologies and tools. Such fixes could be pointless simply because macro-issues are left unaddressed, or because they focus on the wrong conception or leave out important normative dimensions of a conception. Second, it centres too much on how a system behaves, rather than on the politico-legal background that gives us moral reasons to act on the system's output. The question then is how to make AI Ethics normatively satisfying.

Winner's (1987) "political ergonomics" provides a promising vantage point for a method of integration combining a plurality of perspectives. With *political ergonomics*, Winner refers to the 'fit' between technology and society. Each artefact applies to different social relations between citizens and institutions, has a different setting in the physical environment (and therefore a distinct effect on its

sures in place that provide some politico-legal embedding of AI. The AI Act would be an additional one. The source, however, would be distinct: one on a national basis, one on a EU basis. In addition, deontic provenance need not be embedded by legislation. Informal political representatives that are not necessarily 'elected' by means of formal selection procedures but rather by bottom-up audience uptake (think of Dr. Martin Luther King in the context of American segregation or Greta Thunberg for climate change) can provide a source of deontic provenance as well. The link between source and legislation is something to be further explored for a full account of deontic provenance, which is unfortunately beyond the scope of this chapter. We thank Merel Noorman for pressing us on these issues.

surroundings), and so each technology raises different technical problems. Such a triadic view of political, spatial, and technical is essential in order to correctly assess how well technology and society fit together.

Winner (1987, p. 26) applied his approach in the eighties to the development of the card computer, a small computer the size of a credit card that made certain activities and transactions such as shopping or banking move from a physical realm to a digital one. Studies, however, warned against how such digitalized practices "pose a threat to privacy and freedom" and create a world of electronic dossiers in which "persons have no control over the kinds of information routinely collected about them and no control over how such information is used" (Winner, 1987, p. 27).

One response by David Chaum in the 1980s to these concerns was an alternative design of the card computer which was supposed to preserve people's privacy, freedom, and autonomy. Under this alternative approach, privacy is respected by means of "digital pseudonyms" (Chaum, 1985 as cited in Winner, 1987, p. 29), where individuals preserve identifying data such as their names or home address. This method of pseudonyms would allow banks and shops to combat fraud, while the individuals would remain in control over their own personal information.[8]

At first glance, Chaum's card computer seems perfect for preserving individual freedom and privacy, and therefore contributes to democracy overall. However, Winner shows how Chaum's laudable intentions for his alternative version of the card computer lack normative theoretical grounding. As Winner (1987, p. 38) points out, Chaum "assumes that privacy is crucial to freedom and that freedom is essential to democracy", but he remains ambiguous on the distinction between public and private, and what conception of democracy he favours. Indeed, do we want a more Rousseauvian form of direct democracy that favours participation, or rather one more aligned with a republican type where one's civil liberty stands central. Moreover, Winner asks why we need freedom and individual privacy as Chaum's card computer would facilitate, and how they would further a healthy and strong democratic society. As the reader is convincingly shown, it most certainly matters that these questions are answered before the actual development of such a technological artefact. Designing for democracy, indeed, requires a sufficiently fleshed out understanding of democracy to be feasible and desirable.

Similarly, Johnson (1997) implicitly applies such a political ergonomics approach when she questioned the oft-assumed democratic character of the Internet. She shows that "being a democratic technology" is an ideal that is easily misinterpreted when we more broadly analyse how and which values are embedded in the technology. Again, ambiguity and lack of conceptual clarity risk idealizing a technology while missing a more substantial and fundamental impact on society. As Johnson (1997, pp. 24–25) argues, for the internet to be a democratic technology we need to have a perspective on the historical roots of its development, the institutions, practices, and people who maintain the technology. Moreover, we need to know about the materiality of the technology, and the cultural meaning of the values that are

[8] See for further technical elaboration Chaum (1985) and Winner (1987).

embedded. For each of these understandings, our perspective may change regarding the question of whether the technology is democratic. And so, only by considering the larger picture can we prepare a judgment on design requirements.

As we see, both Winner and Johnson recommend combining political theory, history, economics, philosophy of technology, and science and technology studies in order to assess the artefact's political ergonomics, i.e., the fit between either the card computer and society or the internet and society. They approach the technology from a plurality of perspectives and show that depending on how one approaches a technology, also the approach to its design, development, deployment, and regulation will change. They therefore urge us to rethink the artefact itself completely.

Implicitly, the approach taken by Winner and Johnson draws attention to larger societal questions as well as to the provenance of a technological artefact. Part of the solution to the political naivete and the lack of attention to deontic provenance of AI, then, is a 'political ergonomics' of AI that charts the deontic provenance of socio-technical systems. This can provide us with a co-evolutionary framework to adequately assess the fitting interface between AI and institutions that reflects the politico-legal background in which AI is necessarily situated. Such an approach must consider the complex interplay between contextual background, the shaping of AI, and in turn the shaping of society.

For this, integration with different fields like normative political theory, STS, media studies, and political economy are essential. They provide the conceptual tools necessary to navigate large and complex societal issues by zooming out and incorporating questions related to historical background, social power, authority, and legitimacy essential to situate AI systems in their political context. An emphasis on AI's situatedness addresses concerns related to political naivete and helps trace their deontic provenance.

To some extent, we see such integration already. However, specifically the integration of normative political theory has fallen short. For a long while digital ethics has disregarded political theory as a fruitful source of normative guidance, and political theory has omitted digital ethics as a topic worthy of serious reflection (Van den Hoven, 2017[1994]; Gabriel, 2022). Questions related to injustices and violations of human rights to scholars concerned with ideological critique of the digital era were therefore often left to political economy (e.g., Zuboff, 2019) and media studies (e.g., Rushkoff, 2010). Yet like digital ethics, media studies and political economy perspectives miss the conceptual tools of normative political—most relevantly authority and legitimacy—to adequately capture some of these core problems of AI systems. Consequently, despite some broader integration with other fields, questions related to authority and legitimacy have not received the attention they deserve (Lazar, 2023).

Fortunately, political theory increasingly shows more interest in the digital ethics side. To name a few, Van den Hoven (2017[1994]) has built on Rawls' account of justice to inform our understanding of justice in the information society, and Oosterlaken and Van den Hoven (2011) have used Sen and Nussbaum's capability approach to analyse how digital technologies impact our capabilities. Sharon (2021) has applied Michael Walzer's spheres of justice to online platforms, Habermas

(2022) has extended his own theory of structural change of the public sphere to the impact of social media and online platforms, and Susskind (2022) has used the neo-republican conception of freedom as non-domination to argue for a 'digital republic'. Lazar (2023) is developing a theory of communicative justice, and we now even have an introduction to political philosophy of AI (Coeckelbergh, 2022) and a political theory of the digital age (Risse, 2023).

In addition to these more normative political theory related-approaches, other perspectives from critical, feminist, and decolonization studies can be helpful for a political ergonomics of AI (see e.g., Adams, 2021; Mhlambi & Tiribelli, 2023; Waelen, 2022). These theories question the power asymmetry of society at large, thereby necessarily taking on a broader perspective.

Finally, we can think of more practically-oriented approaches. In the policy-domain, for instance, we see a shift in perspective as well when it comes down to AI. The future of work in particular is severely being challenged by the increasing automation thanks to AI. The European group on Ethics (EGE) (2018) has suggested that these changes may require a radical shift in how we organize our society, and advocates a redesign that puts solidarity centre stage. Such a redesign would require us to think about, for example, a universal basic income or about "contributive justice", which suggests that we owe to each other that each can make an appreciated contribution to society.

Other more practically oriented approaches concern our approach to the design of technologies. Here we also see how such practical approaches build on democratic ideals. Think of the work done on design justice by Costanza-Chock (2018), DiSalvo's (2022) book *Design as Democratic Inquiry*, and Forestal's (2022) work on building community in digital environments by focusing on designing for democracy. These works urge us to look beyond merely meeting certain ethical principles, but to design artefacts with and for a community. This is then different from the example mentioned in the introduction of the algorithm aimed to identify to-be-mulched elderly. Rather than designing this system according to fair, accountable, or transparent principles, this system simply would not have been designed if we reasonably consider the communities wishes, needs, and desires.

As Rawls' famously pointed out in his Theory of Justice, we can distinguish between concepts and conceptions of justice. What we all agree on is that AI Ethics has an important role to play, yet there are important differences in ways of how to conceptualize AI Ethics. In this chapter, we have been skeptical of narrow conceptions of AI Ethics and have argued for a broad conception. Broad conceptions question the politico-economic background that informs AI development, deployment, and even ethical assessment. They aim to address political naivete by placing the technology in its larger societal context, and by doing so they contribute to the elucidating picture of AI's deontic provenance. These approaches seem to be aware of the political framework AI systems are embedded in, and precisely *question* this framework as well where necessary. Broad and narrow conceptions need not to exclude each other, but as we saw in Sect. 2, broader conceptions of AI Ethics are underrepresented in the FAccT and AIES literature.

Although the increase in attention in recent years regarding a more comprehensive approach to ethical analysis is promising, more work can and should be done. We must focus on developing systems from a comprehensive and societal broad approach that situates ethical issues in a larger political economy, *cum* ideological assumptions. This may serve to keep us from political blunders and naivete such as fixing filter bubbles in countries with state censorship, arguing that democratic design is the solution in a country run by a dictator, or insisting on privacy-by-design in countries without checks and balances and democratic control over the use of personal data by intelligence agencies. Ideally, more comprehensive meta-level approaches to AI systems will avoid such unfortunate non-solutions and reduce potential for ethics washing. Approaches that fall under broad conceptions of AI Ethics move beyond a focus on morally acceptable *systems* and include a morally acceptable shaping of society in a broader sense. Only with a broad conception of AI Ethics, then, can we escape the normative disappointing framework when we open the Black Box of AI.

Concluding Remarks

In conclusion, what we find upon the efforts of opening the black box of AI as AI Ethics currently does, while not normatively empty, is disappointingly meagre in a normatively relevant sense. It fails to take into account how much the quality of ethical debates is determined by their being embedded in a larger political, economic, and legal system. Just like we need a certain distance to be able to identify what impressionist paintings like those of Monet or Van Gogh depict, we need to approach ethical AI from a certain distance in order to get an adequate impression what moral and political concerns are truly at stake. Otherwise, we risk simple and complex political naivete and lack a clear picture of AI's deontic provenance, without which it is unclear what our reasons are to respect the output of the system. To strive for moral progress should ultimately be the goal of any innovative endeavour, including those in the field of AI. We therefore now turn to our fellow digital ethicists to not (just) ask whether a system's behaviour is 'ethical', but also ask how it is normatively constructed, how it ended up—and with which deontic credentials—at the time and place when and where its output is acted upon, and how the system makes a difference in practice.

References

Adams, R. (2021). Can artificial intelligence be decolonized? *Interdisciplinary Science Reviews, 46*(1–2), 176–197.

Bijker, W. (1995). Sociohistorical technology studies. In S. Jasanoff, G. E. Markle, J. C. Petersen, & T. Pinch (Eds.), *Handbook of science and technology studies*. Sage Publications.

Binns, R. (2018). Fairness in machine learning: Lessons from political philosophy. *In Conference on fairness, accountability and transparency* (pp. 149–159). PMLR.

Birhane, A., Ruane, E., Laurent, T., Brown, M. S., Flowers, J., Ventresque, A., & Dancy, C. L. (2022, June). The forgotten margins of AI ethics. In *Proceedings of the 2022 ACM conference on fairness, accountability, and transparency* (pp. 948–958).

Bolte, L., Vandemeulebroucke, T., & van Wynsberghe, A. (2022). From an ethics of carefulness to an ethics of desirability: Going beyond current ethics approaches to sustainable AI. *Sustainability, 14*(8), 4472.

Brandstedt, E. (2014). An interview with professor Simon Caney. *De Ethica, 1*(1), 71–84.

Burri, T. (2022). The new regulation of the European Union on artificial intelligence: Fuzzy ethics diffuse into domestic law and sideline international law. In *The Cambridge handbook of responsible artificial intelligence*. CUP.

Carnegie Council. (2021, November 10). *Why are we failing at the ethics of AI?* Carnegie Council for Ethics in International Affairs. https://www.carnegiecouncil.org/media/article/why-are-we-failing-at-the-ethics-of-ai

Chaum, D. (1985). Security without identification: Transaction systems to make big brother obsolete. *Communications of the ACM, 28*(10), 1030–1044.

Costanza-Chock, S. (2018). Design justice: Towards an intersectional feminist framework for design theory and practice. *Proceedings of the Design Research Society*.

Crawford, K. (2021). *The atlas of AI: Power, politics, and the planetary costs of artificial intelligence*. Yale University Press.

DiSalvo, C. (2022). *Design as democratic inquiry: Putting experimental civics into practice*. MIT Press.

Dobbe, R., Gilbert, T. K., & Mintz, Y. (2021). Hard choices in artificial intelligence. *Artificial Intelligence*, 300, 103555.

European Commission. (2021, April 21). Lex – 52021PC0206 – en – EUR-lex. EUR. Retrieved May 1, 2023, from https://eur-lex.europa.eu/legal-content/EN/TXT/?uri=celex%3A52021PC0206

European Group on Ethics (EGE). (2018, December 19). Future of work, future of society. *Research and innovation*. https://research-and-innovation.ec.europa.eu/knowledge-publications-tools-and-data/publications/all-publications/future-work-future-society_en#files

Floridi, L., & Cowls, J. (2022). A unified framework of five principles for AI in society. In *Machine learning and the city: Applications in architecture and urban design* (pp. 535–545).

Forestal, J. (2022). *Designing for democracy: How to build community in digital environments*. Oxford University Press.

Gabriel, I. (2022). Toward a theory of justice for artificial intelligence. *Daedalus, 151*(2), 218–231.

Galvagna, C. (2023, March 30). *Discussion paper: Inclusive AI governance*. Ada Lovelace Institute. https://www.adalovelaceinstitute.org/report/inclusive-ai-governance/

Gebru, T. (2022, May 26). *Disrupting big tech – Timnit Gebru*. YouTube. https://www.youtube.com/watch?v=dENwLu1pQb4&t=4s

Gerdes, A. (2022). The tech industry hijacking of the AI ethics research agenda and why we should reclaim it. *Discover Artificial Intelligence, 2*(1), 25.

Gunning, D., Stefik, M., Choi, J., Miller, T., Stumpf, S., & Yang, G. Z. (2019). XAI—Explainable artificial intelligence. *Science Robotics, 4*(37), eaay7120.

Habermas, J. (2022). *Ein neuer Strukturwandel der Öffentlichkeit und die deliberative Politik: Platz 1 der Sachbuchbestenliste der WELT*. Suhrkamp Verlag.

Hagendorff, T. (2020). The ethics of AI ethics: An evaluation of guidelines. *Minds and Machines, 30*(1), 99–120.

Hirota, Y., Nakashima, Y., & Garcia, N. (2022). Gender and racial bias in visual question answering datasets. In *2022 ACM conference on fairness, accountability, and transparency* (pp. 1280–1292). https://doi.org/10.1145/3531146.3533184

Jobin, A., Ienca, M., & Vayena, E. (2019). The global landscape of AI ethics guidelines. *Nature Machine Intelligence, 1*(9), 389–399.

Johnson, D. G. (1997). Is the global information infrastructure a democratic technology? *ACM Sigcas Computers and Society, 27*(3), 20–26.
Johnson, D. G. (2021). Algorithmic accountability in the making. *Social Philosophy and Policy, 38*(2), 111–127.
Keyes, O., Hutson, J., & Durbin, M. (2019, May). A mulching proposal: Analysing and improving an algorithmic system for turning the elderly into high-nutrient slurry. In *Extended abstracts of the 2019 CHI conference on human factors in computing systems* (pp. 1–11).
Laux, J., Wachter, S., & Mittelstadt, B. (2023). *Trustworthy artificial intelligence and the European Union AI act: On the conflation of trustworthiness and acceptability of risk.* Regulation & Governance.
Lazar, S. (2023, January 25). *Lecture I: Governing the algorithmic city*. YouTube. Retrieved April 24, 2023, from https://www.youtube.com/watch?v=MzRWdpB39qw
Lehtonen, M. (2004). The environmental–social interface of sustainable development: capabilities, social capital, institutions. *Ecological Economics, 49*(2), 199–214.
Lukes, S. (2021). *Power: A radical view*. Bloomsbury Publishing.
Maas, J. (2022). A neo-republican critique of AI ethics. *Journal of Responsible Technology, 9*, 100022.
Maas, J. (2023). Machine learning and power relations. *AI & Society, 38*(4), 1493–1500.
Mhlambi, S., & Tiribelli, S. (2023). Decolonizing AI ethics: Relational autonomy as a means to counter AI Harms. *Topoi*, 1–14.
Mittelstadt, B. (2019). Principles alone cannot guarantee ethical AI. *Nature Machine Intelligence, 1*(11), 501–507.
Morozov, E. (2014). *To save everything, click here: Technology, solutionism and the urge to fix problems that don't exist*. Penguin Books.
Oosterlaken, I., & Van den Hoven, J. (2011). ICT and the capability approach. *Ethics and Information Technology, 13*, 65–67.
Perarnaud, C. (2023, April 25). *With the AI act, we need to mind the standards gap – CEPS*. CEPS. https://www.ceps.eu/with-the-ai-act-we-need-to-mind-the-standards-gap/
Perrigo, B. (2023, June 20). *Exclusive: Openai lobbied E.U. to water down AI regulation*. Time. https://time.com/6288245/openai-eu-lobbying-ai-act/
Risse, M. (2023). *Political theory of the digital age: Where artificial intelligence might take us*. Cambridge University Press.
Rushkoff, D. (2010). *Program or be programmed: Ten commands for a digital age*. Or Books.
Schaffer, J. (2016). Grounding in the image of causation. *Philosophical Studies, 173*(1), 49–100.
Schyns, C. (2023, February). *The lobbying ghost in the machine*. Corporate Europe Observatory. https://corporateeurope.org/en/2023/02/lobbying-ghost-machine
Searle, J. R. (1995). *The construction of social reality*. Simon and Schuster.
Selbst, A. D., Boyd, D., Friedler, S. A., Venkatasubramanian, S., & Vertesi, J. (2019, January). Fairness and abstraction in sociotechnical systems. In *Proceedings of the conference on fairness, accountability, and transparency* (pp. 59–68).
Sharon, T. (2021). From hostile worlds to multiple spheres: towards a normative pragmatics of justice for the Googlization of health. *Medicine, Health Care and Philosophy, 24*(3), 315–327.
Susskind, J. (2022). *The digital republic: on freedom and democracy in the 21st century*. Bloomsbury publishing.
Trogdon, K. (2018). Inheritance arguments for fundamentality. In *Reality and its structure: Essays in fundamentality* (pp. 182–198).
Van den Hoven, M. J. (2017[1994]). Towards ethical principles for designing politico-administrative information systems. In *Computer Ethics* (pp. 193–213). Routledge.
Van Maanen, G. (2022). AI ethics, ethics washing, and the need to politicize data ethics. *Digital Society, 1*(2), 9.
Veluwenkamp, H., & van den Hoven, J. (2023). Design for values and conceptual engineering. *Ethics and Information Technology, 25*(1), 2.
Waelen, R. (2022). Why AI ethics is a critical theory. *Philosophy & Technology, 35*(1), 9.

Winner, L. (1980). Do artifacts have politics? *Daedalus*, 121–136.
Winner, L. (1987). Political ergonomics: Technological design and the quality of public life. In *Wissenschaftszentrum Berlin für Sozialforschung* (pp. 87–87). IIUG dp.
Winner, L. (1993). Upon opening the black box and finding it empty: Social constructivism and the philosophy of technology. *Science, Technology, & Human Values, 18*(3), 362–378.
Zuboff, S. (2019). *The age of surveillance capitalism: The fight for the human future at the New Frontier of Power*. Profile Books.

Chapter 6
The Governance of AI Technologies

Merel Noorman

Abstract Norms and values are not stable things. They change over time and in response to, among other things, the introduction of new technologies. This is particularly clear in the developing field of AI Governance, where the introduction of new technologies instigates and triggers renegotiations about the values and norms that these new technologies put under pressure. One element of Deborah Johnson's more recent work is the focus on negotiations around values and norms as part of changing practices following the introduction of computational technologies, like AI. This work aligns with the practice turn in philosophy and the social sciences, where scholars have argued that a focus on practices provides a promising perspective to think about human activity and social life. The focus on practices offers opportunities for empirical philosophical research on how interventions can be made in these negotiations and practices 'in the making'. This paper explores the potential role empirical philosophers and computer ethicists can play in shaping the embedding and governance of AI technologies in practice. It examines this role in the context of AI governance clinics organized for municipalities to help smart city projects address ethical and social issues that go beyond the formal legal and policy questions.

Keywords Practices · AI governance · Ethics · Empirical philosophy

---

M. Noorman (✉)
Tilburg Institute for Law Technology and Society, Tilburg University, Tilburg, The Netherlands
e-mail: m.e.noorman@tilburguniversity.edu

M. Noorman, M. Verdicchio (eds.), *Computer Ethics Across Disciplines*, Philosophy of Engineering and Technology 44,
https://doi.org/10.1007/978-3-031-87974-6_6

Introduction

AI-based applications and services are often presented as technological quick fixes to difficult sociotechnical collective problems in cities, such as crowd control or public health (Sharon, 2021). Although AI certainly has the potential to help address some of these problems, it also introduces a new set of challenges, including the safeguarding of public interests and democratic values in innovation processes (Dameski et al., 2022; Noorman et al., 2022; Taylor et al., 2020). Such data-driven AI-based technologies destabilize existing practices and trigger renegotiations about established norms and values (Lanzing, 2023). An example is the use of data-driven technologies to monitor and track the public in the interests of security or public health, as in the case of crowd monitoring systems or COVID contact-tracing apps, that take shape within extended public-private collaborations (Dameski et al., 2022; Lanzing, 2021). Such initiatives have been accompanied by discussions on, for example, the balance between individual autonomy and collective responsibility, privacy and public health, and the distribution of responsibility. In these discussions, policymakers, researchers, companies, civil society organizations, and many other affected parties negotiate how to understand and weigh moral concepts such as privacy, individual autonomy, and responsibility and how to translate them into acceptable norms, behavior, and new technologies.

Negotiations about norms and values are a recurring element of Deborah Johnson's work. In her contribution to this volume and in earlier work, she argues that values like accountability and responsibility and the norms that translate these values into behavior are not stable but are shaped by ongoing discussions triggered by the introduction of new technologies, such as algorithmic systems (2021). In her analysis, she shows how the current discourse on algorithmic accountability represents negotiations about 'effective accountability practices'. By looking at algorithmic accountability as a social practice, she directs the attention to aspects of these practices under negotiation, such as "the relevant actors to call to account, the forums to which these actors should be expected to account, the beliefs and norms that should constitute the callings to account, when where and how this is done (performed), and the sanctions that can be rendered" (this volume, p. 18).

With the emphasis on accountability as a social practice, Johnson's work aligns with the practice turn in philosophy and the social sciences. Various scholars have argued that practices—understood as arrays of embodied and materially mediated human activity—provide a promising perspective to analyze, study, and think about human life and society in terms of shared skills and understanding constituting social order and change over time (Schatzki et al., 2001). This perspective departs from traditional ways of thinking that focus on, for example, individual minds and actions or systems and structures. This allows for a better understanding of changes in expectations, beliefs, relationships, and power dynamics.

A practice approach can help to make sense of developments in AI governance. AI governance generally refers to the efforts on various levels, ranging from multinational initiatives to organizational-focused approaches, to steer AI development, use, and management in some preferred direction (Gasser and Almeida, 2017; Mäntymäki et al., 2022). The discourse around algorithmic accountability and the discussions about the responsible use of AI for public health crisis response measures are part of this growing field. This relatively new and developing field provides an example of a domain in which negotiations about values and norms are continuously and explicitly taking place, with numerous principles, guidelines, regulations, tools, and approaches being developed. We can see these efforts as part of the negotiations following the destabilization of established moral practices (partially) triggered by the development of new technologies. The focus on such negotiations in emerging and changing AI governance practices provides opportunities for empirical philosophical research to study how practices take shape in interaction with changing values and norms. Moreover, it offers a chance to explore how interventions can be made in these negotiations.

This chapter examines what the role of an empirical philosopher or ethicist can be as AI governance practices develop and are negotiated in smart city innovation projects that center on data-driven digital technologies. The analysis draws on two governance clinics in a European city which consisted of a series of interviews and workshops with the team members of two projects and external experts. Both clinics involved projects that aimed to use data analysis techniques to improve some aspects of crowd management within the city by making use of heterogeneous data before and during the COVID epidemic.

My colleagues and I were asked to help the two project teams think through some of their ethical concerns about aspects of the projects that they were working on. This presented an opportunity for us to study how negotiations about values and norms take shape within existing governance practices, and what kind of interventions could be made in these negotiations. Our ambition for these clinics was to go beyond identifying and weighing several ethical values—there were already plenty to go around—and translating them into designs. Rather, we were interested in whether and how these projects would be able to address ethical issues and concerns within the existing governance practices and how technologies in development challenged existing governance structures. The clinics did not focus strictly speaking on AI technologies. Rather, they centered on data-driven systems and what could potentially be done with various data sources through AI and other data analytic means. These projects nevertheless provided a good setting to explore the role of practice-oriented philosophy in AI governance. They showed how, when, and what ethical issues in the development of data-driven technologies surfaced, and how values, norms, and responsibilities in decision-making processes were negotiated as part of the changing governance structures around these projects.

In the following, I first outline the growing field of AI governance. Subsequently, I argue that a practice approach to studying the developments in this field can help to understand how and when moral values and norms can become subject to debate, and how this destabilization is part of negotiations about how to constitute 'effective governance practices'. I then show, based on the work that my colleagues and I did in the clinics, how this approach can help to understand organizational governance practices ‘in the making’. Finally, I argue that there is a role to play for empirically oriented philosophers in shaping these governance practices.

AI Governance

With the increasing pervasiveness of AI technologies come growing concerns about their possible negative effects. In response to these concerns, governmental and non-governmental actors have proposed a wide range of ethical guidelines, manifestos, principles, as well as legislation to steer developments in AI in more positive directions and to minimize harms resulting from the use of these systems (Fjeld et al., 2020). These initiatives are part of the growing field of AI governance, which in turn is part of a broader governance landscape, including corporate governance, data governance, and governance of the digital (Mäntymäki et al., 2022). An unambiguous commonly accepted definition of AI governance is, nevertheless, lacking and literature on the topic has focused on a diversity of activities including, the development of international ethical guidelines, legislation and human rights, organizational processes, and procedures, as well as technical tools to facilitate accountability, transparency, and explainability of AI technologies (Cath, 2018; Dafoe, 2018; Smuha, 2021; Taeihagh, 2021).

To characterize the field, some scholars have attempted to categorize different layers or overlapping domains, including social, legal, ethical, organizational, and technical layers of AI governance (Gasser & Almeida, 2017; Shneiderman, 2020). Luciano Floridi in his discussion of digital governance, closely related to AI governance, offers a helpful categorization in this regard between regulation, ethics, and governance. He describes governance as one of three normative forces that can shape how we decide and coordinate what we want the future of digital technology and human society to look like (2018, p. 4). The other two are digital ethics and digital regulation. Digital ethics, according to Floridi, is the branch of ethics that studies and evaluates moral problems that are connected to data, information, algorithms, and corresponding practices and infrastructures. Through moral evaluations, it shapes digital governance and regulation. Digital regulation is the system of rules that regulate the behavior of relevant agents, elaborated, and enforced by social and governmental institutions. Digital governance, he argues, is “the practice of establishing and implementing policies, procedures, and standards for the proper development, use and management” of the digitally infused world we live in (2018, p. 3). Governance is both the processes and procedures as well as the overarching force that coordinates between digital ethics and digital regulation, he states. They are, thus, interdependent forces.

Absent a commonly accepted overarching definition, I will consider AI governance in this chapter to be a concept that refers to activities that aim to steer developments regarding AI in a preferable, responsible, or ethical trajectory at various levels of organization, from global multilateral negotiations to local organizational development protocols. It goes beyond the strictly governmental and legal structures that set the terms for what can and cannot be done and includes the informal expectations and technical norms that guide behavior. AI governance encompasses the whole range of institutions, agencies, and groups that have a stake in setting boundaries for a particular action or project. Moreover, it points to the mechanisms, strategies, structures, power relationships, technologies, discourses, and norms that guide the conduct of people, organizations, and countries. Simply put, AI governance asks how to organize the development, deployment, and use of AI technologies in such a way that leads us toward explicit goals. In Europe, for example, such goals can include economic growth as well as responsible and trustworthy technological development.

What the target of this type of governance entails is more difficult to define, as this is very much still a topic of ongoing discussion. The European Union has developed its definition of AI, in interaction with other organizations, such as the OECD (Bertuzzi, 2023; European Commission, 2023). Generally speaking, according to these organizations, AI refers to machine-based systems that "for explicit or implicit objectives, infers, from the input it receives, how to generate outputs such as predictions, content, recommendations, or decisions that can influence physical or virtual environments" (Bertuzzi, 2023). Other definitions refer to advanced or future forms of AI technologies when discussing AI governance (Dafoe, 2018; Maas, 2023), while yet others use AI interchangeably with algorithms, automated decision-making, or autonomous systems (Mökander et al., 2023). For this chapter, a precise definition is not required as the focus is on governance practices and how they change in interaction with the development and introduction of new digital technologies.

One aspect of AI governance that has received considerable attention has been the formulation of ethical values, principles, and guidelines (Fjeld et al., 2020). Through defining sets of core values and principles, regulators, non-governmental organizations, and industry players have outlined frameworks intended to give direction to AI development and implementation trajectories. We can see these moral evaluation efforts as informing a mode of governance that takes the shape of norm-setting through rhetorical strategies and prescriptive guidelines (Floridi, 2018). Thus, insisting on abstract values such as transparency, accountability, fairness, etc. informs the shaping of consensus about the norms that AI should adhere to. However, these norm-setting activities are only a part of AI governance.

The operationalization of ethical principles and guidelines is another part of AI governance that only recently received more attention as organizations and companies struggled to translate abstract principles into practice. Several initiatives have developed ethical tools and methodologies to identify and define key values in particular projects, to then sketch design options that would promote these values more than other values (Bleher & Braun, 2023; Verbeek & Tijink, 2020). Such approaches build on earlier approaches in computer ethics and other fields, such as value-sensitive design, participatory design, and responsible innovation (Ryan & Blok, 2023; Shneiderman, 2020). Other initiatives have developed technical tools that help developers to monitor alignment or compliance with particular principles, guidelines or regulation (Morley et al., 2020).

The critique of the first wave of guidelines and their operationalization was that they tended to be toothless (e.g. Hagendorff, 2020; Mittelstadt, 2019). Although there are incentives for companies and others to follow such guidelines—such as being perceived as 'good'—there is no obligation. Critics of principle-based approaches have consequently argued for regulation combined with enforcement. In recent years, we have seen more initial steps to regulate AI, including the drafting of the European AI act and the US White House's Executive Order on safe, secure, and trustworthy AI. Such critiques and regulatory initiatives give further shape to AI governance.

In the following, however, I will concentrate on negotiations at the intersection of ethics and governance in relation to AI. This is where I see a clear role for empirically oriented philosophers because AI governance is still very much in the early phases of development, and existing practices and their underlying values and norms are still actively negotiated. Moreover, experimental AI technologies are intended to push the boundaries of what is possible or known. They are developed to do things that human beings could not do before and as a result, disrupt and destabilize existing ways of doing things. With that, they change expectations, challenge established norms, and shift roles and responsibilities (Danaher & Sætra, 2023; Nickel et al., 2022; Swierstra et al., 2009). The recent rise of foundational models and large language models provides a vivid example of this. These technologies have in a very short period destabilized existing practices in many domains of life, from work to education to our personal lives (Hopster & Maas, 2023). They have prompted discussion about the way we relate to others, perform tasks, or evaluate actions and our environment. Should legal advisers be able to make use of generative AI to write legal briefs? Should lecturers assess students on their prompt engineering skills rather than their provision in writing essays? Can commercial companies use material collected from social media to train generative models?

The development and implementation of AI technology is a moment where ethical questions emerge; questions about whether core values are still safeguarded in the use of AI technology, how we prioritize values, or whether established norms of behavior still suffice. These are questions that have no ready answer. They require further reflection and discussion. We have seen these negotiations happen on global and national levels in recent years, with the development of a wide range of manifestos, guidelines, principles, and new regulations, as mentioned before. Moreover, a growing number of initiatives aim to operationalize these high-level abstract principles on a more local level and integrate them into existing practices, through tools, standards, training, and more. Developers, project managers, policymakers, and others involved are learning on the fly and through ongoing negotiations in development and implementation processes.

There is little ethical research that goes beyond the initial point of identifying values and design options and looks at shaping the governance of AI in practice. Research on how these AI systems should be developed and embedded in existing practices and existing power landscapes is only just coming off the ground. This is problematic because many of the current developments take place in experimental and agile settings, in which the distinctions between production and use of these systems have collapsed and normative questions continuously arise (Gürses & Van Hoboken, 2017). Digital data-driven systems are increasingly being developed as part of ongoing iterative building and testing processes, as if they are in a perpetual prototyping stage. Design requirements may change is in relatively short time, which makes performing ethical analyses and impact assessments ahead of development difficult. Yet, in these experimental development processes, decisions are made that often have a moral dimension and it is not always clear who should make these decisions, based on what information or how they should be made. One-off ethics exercises or impact assessments as part of governance are therefore not enough; AI governance requires continuous ethical reflection through all stages of a systems lifecycle.

Giving shape to AI governance requires a closer look at practices and how negotiations take place between the different actors and in interaction with the technology and the environment. The focus in this chapter will therefore be on AI governance at the level of organizations. As I will argue in the next section, philosophical research that actively engages with governance practices 'in the making' can provide a valuable perspective on the interactions between assumptions about the technologies, the context in which they will be used, the power relations that they reify or challenge, the human beings affected by them, and the values and norms that these technologies are supposed to reflect.

Practice-Oriented Philosophical Research

Practice theory as elaborated in science and technology studies can help to further flesh out the concept of changing practices and the role of empirical philosophers and ethicists in it, particularly in the context of technological development. The notion of practices in the practice theory tradition refers to routinized forms of human activity centered on shared and embodied understanding (Schatzki et al., 2001). Although practice theory refers to a collection of different approaches to thinking about and studying social order and human activity, these approaches share several commonalities (Reckwitz, 2002). In particular, in these theories, the concept of practice provides an analytical focus that extends beyond the individual to encompass a constellation of bodies, minds, and things. Practices are the arrays of routinized activities in which these various components are inextricably linked. These activities are constituted by and reinforce the shared and practical understanding that human beings need to perform their activities.

In practice theory, activities are thought of as constitutive of as well as reproduced by shared and embodied understanding (Schatzki et al., 2001). This is the kind of practical understanding that people need to perform meaningful activities. Shared understanding emphasizes the social aspect of practices: people can perform their activities because they interact with other people and are part of a group that over time develops some shared routinized behaviors that are meaningful to them. The practice of playing chess consists of a set of routinized activities that are meaningful to those people who have a shared and practical understanding of what chess is about. To some extent, this shared understanding can be written down in rules (the rules of playing chess), but a large part of this understanding takes the shape of embodied skills and tacit knowledge that people acquire through their socialization and their interactions with their environment. Understanding is, thus, not just about knowing *that,* but also about knowing *how.* People can perform certain routinized activities because they have the knowledge and skills to do an activity in a meaningful way. Their bodies have learned to perform an action, just as they have learned the rules of the game. The mental aspect of understanding, according to Schatzki, relates to more than knowledge; it also has an emotional or motivational aspect to it. People carry out a practice not only in terms of knowledge and skills but also in terms of associating feelings and goals with these activities. Understanding then is reflected in the bodily movements, goals, and motivations.

As the chess example shows, another key component of practices is 'things', or artifacts and objects (Reckwitz, 2002). Human beings perform activities with and through artifacts and objects, which gain their meaning through these routinized activities. How a mobile phone is used, the norms associated with this use, and the status that users derive from it are an effect of practices. At the same time, the way we communicate over long distances cannot be separated from these artifacts. To understand this kind of communication we must look at what behavior the artifacts afford and how they reflect assumptions about communication. Things, thus, reproduce the social.

Things can be material, like chess pieces, footballs, or smartphones, but they can also take the form of language or symbols. Values such as 'respect' gain their meaning through the use of language within particular practices, while they simultaneously reify these practices (Anker & Afdal, 2018). By expressing expectations about respect, a shared understanding develops. However, the meaning of this value also takes shape and is expressed through the material and symbolic set-up of practices. As Reckwitz, puts it: "[a] practice is thus a routinized way in which bodies are moved, objects are handled, subjects are treated, things are described, and the world is understood" (Reckwitz, 2002, p. 250).

Practices embody and reinforce shared, historically-constituted structures, but they are not immutable (Schatzki et al., 2001). They continuously change. Nicolini, therefore, helpfully emphasizes the performative aspect of practices by conceiving of them as regimes of performances of organized saying and doings that "have a history, social constituency, and a perceivable normative dimension" (2017, p. 5). They exist as long as they are performed and reproduced by a social group that legitimates and gives meaning to them. Although there are structural elements of practices, such as institutions, these should be considered the effect of such routinized social practices. This means that these structures can change in response to breakdowns in routines (Reckwitz, 2002).

Routines are continuously challenged and negotiated for various reasons. One such reason is the introduction of new technologies. The introduction of new technology destabilizes established ways of doing things. They enable human beings to do things differently, more efficiently or accurately. They create or enable new kinds of activities and knowledge and constrain and inhibit others. At the same time, this also introduces indeterminacy and uncertainty. This requires the establishment of new routines. Human beings must figure out, learn, and agree upon what to expect from the technology and others in relation to that technology and the activities that it enables.

Changes in practices are therefore accompanied by negotiations. Negotiations about what things mean and how important they are. Johnson shows how taking a closer look at such practices gives a richer insight into how concepts, like accountability, develop in the interplay between technology and context (Johnson, 2021). Accountability practices from this perspective are the routinized ways that people understand, evaluate, ascribe, and enact accountability, as Johnson explains. They are constellations of interconnected elements including actors, publics, knowledge, skills, formal and informal rules, documents, technologies, bodily movements, and institutions. They encompass behaviors such as accepted ways of evaluating actions, holding others to account, blaming, or praising, and conveying expectations about obligations and duties, which may be articulated and reified through instruction manuals, ethical codes, observation of past practices, training programs, directives, policy documents, laws, and more (Noorman & Johnson, 2014). When digital technologies are introduced that displace existing roles and responsibilities, the various elements that make up the accountability practices are renegotiated: expectations about what human actors can do and know may change, relationships between actors shift, evaluation criteria may be up for discussion, etc.

Similarly, governance practices are not stable, but are continuously negotiated. These practices are the practices that assign responsibilities and accountabilities and guide power relations. They consist of expectations about who is responsible for what, mechanisms for holding people to account, ethical codes, concepts, protocols, documents, and more. With the introduction of new technologies, such as AI, these practices are put under pressure and new negotiations about how to steer activities within a particular domain take place. We can see the discussions around AI governance, such as those ethical principles or new legislation, as part of these negotiations.

A practice-oriented approach provides, thus, an analytical lens to follow these negotiations as they take shape and potentially make interventions in these negotiations. This approach focuses our attention on the actors involved, their routinized behaviors, expectations, and shared understanding, as well as all the tools, documents, concepts, motivations, and technologies used to steer this behavior and knowledge-making activity. It also directs attention to how ethical concepts are interpreted, appropriated, and applied to give shape to these negotiations. A practice-oriented approach to philosophical research, such as Johnson's analysis of accountability, involves looking closer at practices to understand how accountability and other values are negotiated and how interventions can be made.

In the next section, I will further explore the role of empirical philosophy in governance practices 'in the making' in the context of an illustrative example where the development and introduction of data-driven technologies set off local negotiations about how to ensure the responsible development and use of the technology.

The Role of Empirical Philosophy in AI Governance

Philosophy and ethics are part of negotiations about AI governance. Theories and philosophical discussions inform negotiations about which values to prioritize or how best to develop and implement these technologies (Gal, 2019; Hagendorff, 2020; Mhlambi, 2020). Drafters of ethical guidelines and their critics look towards ethical theories, such as utilitarianism, deontology, or ubuntu, to provide foundations for selected principles or to critique them. However, philosophers and ethicists can play a more active role in these negotiations. Governance of AI, currently, applies to situations in which the technologies are not fully developed, the applicability of existing norms and rules is uncertain, and those who might eventually be affected may not yet have had much concrete experience with how particular AI technologies change established practices (Noorman & Johnson, 2014). This requires governance practices that are sensitive and responsive to disruptive and norms-pushing effects and that can deal with continuous questions and ethical deliberations about what is permissible, acceptable, or in support of the good life. Philosophers and ethicists can on the one hand contribute to these negotiations by developing a better understanding of these negotiations through descriptive conceptual analyses, and on the other hand through direct interventions in negotiations taking place in different governance practices based on these analyses.

To illustrate the potential of the two kinds of philosophical research on governance practices around the development of AI-based or data-driven technologies, I will briefly discuss two governance clinics that we ran in a European city (Jameson et al., 2021). We were asked to collaborate with two projects focused on crowd control to help think through how they could responsibly develop these technologies. The city's policymakers and administrators were aware of the wide range of issues that come with the new data-driven technologies they hoped to deploy and had taken several steps to address some of these concerns. They were in the early stage of adapting the existing governance structures through several top-down initiatives, such as the adoption of a set of ethical principles, a city-wide data strategy, a digital agenda, and more. Yet, these high-level abstract policy instruments did not easily translate to the practice of designing, developing, and deploying new innovative technologies, as became clear during the clinics. Although aware of the city's initiatives for responsible data governance and AI development, the project team members who consulted us were unsure of how to, for example, interpret the ethical principles or who was authorized to make decisions in case of uncertainty about the right interpretation. To help the team members address such questions we organized governance clinics which consisted of several preparatory interviews and two or three consecutive workshops with team members, to work with them in identifying gaps in their current governance structures and formulating ways of filling these gaps. A key part of the workshop was a guided discussion about concerns the team members encountered.

The dynamics of the two projects provided a rich setting for analyzing how negotiations about governance practices around new digital technologies take place and to potentially facilitate or even steer them. Both through descriptive analysis and through several interventions, we worked with the project team to help them develop governance practices that aligned with the city's objectives regarding responsible development and use of data-driven and AI-based systems. In the following, I will briefly discuss the two approaches that philosophers can take drawing on the clinics.

Descriptive Analyses of Governance Practices

A practice approach to philosophical research can take the form of empirically informed descriptive analyses of changing governance practices in particular environments. This can provide a deeper understanding of how values and norms, including those featured in the various AI governance policy documents, are destabilized, renegotiated, and institutionalized with the introduction of new technologies. This approach draws attention to routines and where they break down, where shared knowledge is questioned, and how relationships are maintained or become unstable. Moreover, it sheds light on how technologies and even discussions about future technologies have a performative role in conceptualizing values and norms (Danaher & Sætra, 2023).

Negotiations about a shared understanding of values and norms were central to the governance clinics, triggered by technical design decisions and shaped by conflicting expectations. In one of the clinics, for example, the interpretation of values was a recurrent theme and point of friction in the discussions around the new technologies. The city had adopted a manifesto that outlined its vision of a responsible digital city. The six key values that this manifesto listed, such as ownership and control, turned out to be less than self-evident. The use of commercial data to provide a more fine-grained insight into the movements of crowds and to potentially model these movements raised questions about whether this aligned with the city's ethical values. The city's manifesto expressed the value of 'control' in terms of the principle that citizens should have control over data to increase the influence and freedom of the citizens. However, it was unclear to the team members how far this control should reach. Could commercial data only be used if citizens had consented to each particular use, for example? Or would blanket consent be sufficient? Moreover, the use of commercial data would most likely create a dependency on commercial firms, as the accuracy of the systems would be compromised if the data were no longer available.

At the same time, the team members felt pressure to quickly develop a crowd-monitoring system that could compete with Big Tech services. Not using commercial data would slow them down considerably. The municipality was the driving force and financier of the project, but the team leaders aimed to adopt a more corporate approach to the development of a system. This involved an agile development process, where short iterative sprints were made to quickly produce a prototype that could be tested, validated, and further fine-tuned. With this approach also came an emphasis on speed of development, which sometimes clashed with the slower bureaucratic processes of a public organization intended to safeguard the public interest. Commercial data was readily available and could illustrate the potential of a public crowd-management management system. The alternative of asking citizens to volunteer their data was much more time-consuming. To find a way out of this dilemma, the team member negotiated the conditions under which such commercial data could be used while ensuring citizen control over data, such as transparency about the origins of the data.

Our approach to the clinics was to not take a moral position on how values should be interpreted. We, as academic scholars, after all were not the right people to prescribe a particular interpretation. Rather, we were interested in exploring how discussions and decision-making about the values could take shape within the project and the broader organization, and how this was influenced by the different power relations at play. The conversations that took place in the workshops provided insights into the activities and shared understandings that gave shape to what the team members understood as public interests and how implicit assumptions were destabilized. They showed that different practices were at play in these project teams. The negotiations about how to interpret values provided an example of this. The flex workers with a corporate background, hired to develop these technologies, had a different set of expectations about what was required of the system and the system's development than the civil servants. This came out, for example, in disagreements and muted frustration between the privacy officer and some of the developers about what data could be used and under what conditions.

The focus on negotiations about existing and emerging practices also provided a look into the redistribution of roles and responsibilities within existing governance structures. For example, although, there were data protection and privacy protocols and oversight authorities in place, these did not give a straightforward answer to the questions of the project teams, such as questions about how open data should be. These questions extended beyond compliance frameworks and existing governance structures. Moreover, it was unclear who should decide on design options that had a particular political dimension, such as whether to use commercial data or not. The city council and executive management seemed to be too far removed from such detailed technical concerns, while the project teams themselves did not feel like they had a clear mandate to make decisions based on ethical or political considerations. Their performance would, for example, not be evaluated on whether they decided to take more time in the project to not be reliant on commercial data. Implicit in such discussions was the role of the municipality in safeguarding the public interest. As one team member in one of the projects stated, 'There is no process for complying with the notion of public interest.' Ensuring that the public interest is safeguarded in practice means that it is not always clear who is responsible or whose job it is.

In sum, negotiations about governance practices give insight into how values acquire meaning in a specific context and how these translate into norms with their associated roles and responsibilities. Empirical philosophers can critically analyze these meanings against the background of broader reflections on these values.

Interventions in Negotiations

Beyond descriptive and conceptual analyses, philosophers and ethicists can play a more interventionist role by directly participating in negotiations. Participation can be in the form of asking questions, providing different perspectives, or elucidating underlying assumptions and different interests. One example of philosophical research that explicitly engages with practices is Field philosophy (Brister & Frodeman, 2020). Field philosophy is an area of research that aims to bring philosophy to the shop floor. It departs from theoretical philosophy as it aims to engage with the real world and all its complexities. Rather than ideal theory, intuitions, and thought experiments, it examines philosophical issues in their natural habitat in interactions with the people in this habitat. This is based on the idea that only by being in the real world can we understand the messiness and the exemptions to the rule and can we better see what is not intuitive or less visible, such as the effects of social systems on philosophical concepts. Field philosophy does not commit to particular methods but makes use of a wide range of philosophical methods such as phenomenology, participatory observation, and computer modeling. In contrast to other 'disinterested' qualitative social sciences and humanities, such as geography, the field philosophers do not just observe but bring understanding and techniques to use. They actively engage and collaborate with people on the shop floor to work through particular problems. They do not start from theoretical abstraction but respond to the specifics of the actual situations related to philosophical concerns, such as ethics, justice, and aesthetics. "Field philosophers embed themselves in natural contexts, clarifying goals and values and expanding the decision space for complex problems" (Brister & Frodeman, 2020).

In the governance clinics, we adopted a field philosophical approach combined with a focus on governance practices. The aim was to work with the team members to explore the ethical questions that came up in the daily development activities of the team rather than translating a particular set of values or principles. Several, sometimes competing, normative frameworks were relevant to the work the teams did, but it was not always clear how they should relate to or navigate them. By highlighting how existing governance practices were negotiated and explicating some of the assumptions that the teams were basing their arguments on, we aimed to inform their discussions. We, for instance, invited external experts to offer different perspectives on what the team was building by asking questions, such as who is the user of the system, what does the system do and for whom, whose interests are served by the system, and how would one be able to break or game the system? This allowed us, through a guided discussion, to further interrogate the teams' assumptions about, for example, affected groups or users of the system or what the possible unintended consequences of the system could be. By highlighting aspects of the municipality's bureaucratic and democratic practices and their underlying values we asked the project team members to consider how this would affect their ideas about what the crowd-monitoring system was supposed to do and for whom. This exercise provided the teams with a richer understanding of the system and how and by whom it was envisioned to be part of the city. It also provided insights into who was left out or possibly misrepresented in these visions. This, for instance, informed the shared understanding of the multifaceted character of open data: in one project all the team members came to understand that certain forms of open data about the city would primarily benefit companies with the time and resources to work with the data.

In a subsequent step, we then explored with the teams how identified issues could be addressed in the existing governance structures and where the gaps were in these structures. Focusing on one particular issue (e.g. open data), we asked them to think about who should make what decision, and what would be the conditions under which decisions were delegated to others. It became clear, for instance, during these discussions that systematic and explicit ethical reflection was not an integral part of the development practices. When ethical questions were raised, discussions tended to be informal, and questions were often not recognized as ethical. Decision-making about such issues was not explicitly assigned to an actor in the process. Together with the project teams we then explored options to address such gaps.

So why should philosophers take on this role? There are already many empirical research methodologies that examine and interact with all sorts of practices and power relations, such as participatory action research, design, and ethnography. Moreover, there is a growing number of consultants that offers a range of approaches, methodologies, and tools to operationalize AI governance. Philosophers however can bring in a complementary skill set that focuses on making different ethical perspectives and choices explicit and outlining the structure of argumentation, as well as enriching participants understanding of values and the interdependencies between different values. Through interventions in the form of questions during guided discussions, philosophers can create space for such reflection. The clinics showed that ethical questions continuously arise in the development of these technologies. Questions, for example, about what a particular value, such as openness or control, means and how to interpret it. Such questions do not have a ready-made answer and require further negotiation between relevant stakeholders taking the context into account.

Part of what distinguishes a field philosophical approach from a consultancy approach is the positionality of the researcher. There are various expectations of what outside groups do for city projects. In our experience, project teams or organizations tend to ask for ready-made answers to concerns or questions that they have. In the clinics, we purposefully did not want to provide such answers, because our intention with these clinics was to focus on developing governance practices in light of emerging technologies. We looked at how interventions in those practices could be made such that the organization could address similar future questions and concerns appropriately. This required a more Socratic stance: asking questions about governance structures, understandings, roles, and responsibilities, while supporting the learning process of the project team, rather than providing answers. We found, for instance, that in the case of data analytics for municipal services, there is often a much broader vision of risk required than the usual issue of personal data under the GDPR. However, there is no structural process for this, and there is language missing around what it means. The role of the philosophers here is to help in developing that language. Where the discussion remains focused on translating ethical guidelines, projects will work with a much more limited understanding, reduced to ethical checklists and compliance. Our relatively autonomous position as university researchers afforded us to be more Socratic and ask questions that could be uncomfortable or frustrating. This positionality allows for a deeper probing of the practices under negotiations, as it is through these frustrations that practices become visible. Finding ways to resonate with the group's shared understandings is one way to make interventions in ongoing negotiations possible, but creating friction to move forward is necessary at times.

The focus on governance practices also distinguishes, the described approach from ethical tools that focus on eliciting values and developing design perspectives. As mentioned, we did not look for ways for the project in question to comply with a pre-determined set of ethical principles or compliance requirements. Through the discussions in the clinics, several ethical perspectives were made explicit. The aim of our approach was not to arbitrate between these different views. Instead, we worked from the assumption that cities needed to decide for themselves. What the governance clinic approach offers instead is a reflection on the shape of governance architectures and the arrangements of power within them, which determine whose ethics get foregrounded and what power over people in the city accompanies that foregrounding. The aim was to increase recognition and representation of plural interests within these architectures. This, nevertheless, has a normative impact in the sense that it, for example, foregrounds many interests, instead of just economic growth and privatization.

Philosophers can only do so much though in these negotiations. In the case of these new technologies, it remains key that various affected groups and external experts are consulted to bring in different perspectives and identify blind spots. Moreover, a governance clinic is not guaranteed to produce change, but it does produce knowledge and often a will to do things differently. What distinguishes the governance clinic approach is that by placing the focus on practices and how they are renegotiated we can go up a level of abstraction to ask how governance has been conceptualized in a particular project and whether it protects and advances the municipality's and the city's objectives, rather than the project's or those of individuals stakeholders interests. Philosophers can play a role here, in eliciting and unpacking concerns and helping to think through the responsibilities in the governance structure. They can help to clarify goals and values and expand the decision space for complex problems such as data-driven crowd management.

Conclusion

Empirical philosophy has in recent decades shed more light on the situatedness and dynamics of ethics in practice. It has contributed to our understanding of moral values and norms and how they change over time and in interacting with technological development. Practice theory as developed within STS provides a complementary perspective on these analyses and can help bring to light the various components that are involved in the mutual shaping of morality. Johnson's work on algorithmic accountability, for example, shows how analyses at the intersection of empirical philosophy and STS can provide a richer understanding of moral concepts such as accountability and how they are negotiated and reinterpreted in interaction with digital technologies.

In this chapter, I have further unpacked practice theory to discuss on the one hand how it can contribute to advancing the understanding of AI governance as a field in development and the mutual shaping of values, norms, technology, and governance within this field. A practice approach directs attention to the different interlinked components that shape how values and norms acquire meaning, including the actors involved, their routinized behaviors, expectations, and shared understanding, as well as all the tools, documents, concepts, motivations, and technologies. It provides an analytical lens to study how abstract values become understood and enacted in local practices. As such it gives insight into AI governance 'in the making'.

On the other, I argued that empirical philosophers can play an active role in the development of AI governance through practice-oriented interventions in ongoing negotiations. The analysis showed that governance is more than compliance or lists of values and guidelines. It also involves the active negotiation about the meaning and significance of these values and guidelines in particular contingent contexts and about the kinds of institutions, decision-making processes, and power relations that are required to ensure that these values are safeguarded. Here, empirical philosophers can provide a valuable contribution by explicating underlying assumptions and clarifying possible conflicts between different perspectives.

Acknowledgments This chapter is based on research that I conducted together with my colleagues Prof. Linnet Taylor and Shaz Jameson. I am grateful for their insights and for the inspiring collaboration that helped shape many of the ideas in this chapter.

References

Anker, T., & Afdal, G. (2018). Relocating respect and tolerance: A practice approach in empirical philosophy. *Journal of Moral Education, 47*(1), 48–62.

Bertuzzi, L. (2023, September 11). *OECD updates definition of Artificial Intelligence 'to inform EU's AI Act'*. Euractiv. https://www.euractiv.com/section/artificial-intelligence/news/oecd-updates-definition-of-artificial-intelligence-to-inform-eus-ai-act/

Bleher, H., & Braun, M. (2023). Reflections on putting AI ethics into practice: How three AI ethics approaches conceptualize theory and practice. *Science and Engineering Ethics, 29*(3), 21.

Brister, E., & Frodeman, R. (2020). *A guide to field philosophy: Case studies and practical strategies*. Routledge.
Cath, C. (2018). Governing artificial intelligence: Ethical, legal and technical opportunities and challenges. *Philosophical Transactions of the Royal Society A: Mathematical, Physical and Engineering Sciences, 376*(2133), 20180080. https://doi.org/10.1098/rsta.2018.0080
Dafoe, A. (2018). AI governance: A research agenda. In *Governance of AI program* (Vol. 1442, p. 1443). Future of Humanity Institute, University of Oxford.
Dameski, A., Spahn, A., Pouw, C. A., Corbetta, A., Toschi, F., & Bombaerts, G. (2022). How to balance individual and collective values after COVID-19? Ethical reflections on crowd management at Dutch train stations. In *Values for a post-pandemic future* (pp. 215–232). Springer.
Danaher, J., & Sætra, H. S. (2023). Mechanisms of techno-moral change: A taxonomy and overview. *Ethical Theory and Moral Practice, 26*, 1–22.
European Commission. (2023, December 9). *Commission welcomes political agreement on artificial intelligence act**. European Commission Press Corner. https://ec.europa.eu/commission/presscorner/detail/en/ip_23_6473.
Fjeld, J., Achten, N., Hilligoss, H., Nagy, A., & Srikumar, M. (2020). Principled artificial intelligence: Mapping consensus in ethical and rights-based approaches to principles for AI. Berkman Klein Center Research Publication, (2020–1).
Floridi, L. (2018). Soft ethics and the governance of the digital. *Philosophy & Technology, 31*, 1–8.
Gal, D. (2019). Perspectives and approaches in AI ethics: East Asia. In *Oxford handbook of ethics of artificial intelligence*. Oxford University Press. (forthcoming).
Gasser, U., & Almeida, V. A. (2017). A layered model for AI governance. *IEEE Internet Computing, 21*(6), 58–62.
Gürses, S., & Van Hoboken, J. (2017). Privacy after the agile turn. In *Cambridge handbook of consumer privacy*. Cambridge University Press. https://osf.io/preprints/socarxiv/9gy73/ or https://osf.io/ufdvb/
Hagendorff, T. (2020). The ethics of AI ethics: An evaluation of guidelines. *Minds and Machines, 30*(1), 99–120.
Hopster, J. K., & Maas, M. M. (2023). The technology triad: Disruptive AI, regulatory gaps and value change. *AI and Ethics, 4*, 1–19.
Jameson, S., Taylor, L., & Noorman, M. (2021). *Data governance clinics*.
Johnson, D. G. (2021). Algorithmic accountability in the making. *Social Philosophy and Policy, 38*(2), 111–127.
Lanzing, M. (2021). Contact tracing apps: An ethical roadmap. *Ethics and Information Technology, 23*, 87–90.
Lanzing, M. (2023). Traveling technology and perverted logics: Conceptualizing Palantir's expansion into health as sphere transgression. *Information, Communication & Society*, 1–17.
Maas, M. M. (2023). *Concepts in advanced AI governance: A literature review of key terms and definitions*.
Mäntymäki, M., Minkkinen, M., Birkstedt, T., & Viljanen, M. (2022). Defining organizational AI governance. *AI and Ethics, 2*(4), 603–609.
Mhlambi, S. (2020). From rationality to relationality: Ubuntu as an ethical and human rights framework for artificial intelligence governance. *Carr Center for Human Rights Policy Discussion Paper Series, 9*, 31.
Mittelstadt, B. (2019). Principles alone cannot guarantee ethical AI. *Nature Machine Intelligence, 1*(11), 501–507.
Mökander, J., Sheth, M., Watson, D. S., & Floridi, L. (2023). The switch, the ladder, and the matrix: Models for classifying AI systems. *Minds and Machines, 33*(1), 221–248.
Morley, J., Floridi, L., Kinsey, L., & Elhalal, A. (2020). From what to how: An initial review of publicly available AI ethics tools, methods and research to translate principles into practices. *Science and Engineering Ethics, 26*(4), 2141–2168.
Nickel, P. J., Kudina, O., & Van de Poel, I. (2022). Moral uncertainty in technomoral change: Bridging the explanatory gap. *Perspectives on Science, 30*(2), 260–283.

Nicolini, D. (2017). Practice theory as a package of theory, method and vocabulary: Affordances and limitations. *Methodological Reflections on Practice Oriented Theories*, 19–34.

Noorman, M., & Johnson, D. G. (2014). Negotiating autonomy and responsibility in military robots. *Ethics and Information Technology, 16*(1), 51–62.

Noorman, M., Turel, T., Taylor, L., Tol, Y., Groothoff, B., & Pek, J. (2022). *Aanbevelingen en follow-up naar aanleiding van de IGOR datagovernance clinics*.

Reckwitz, A. (2002). Toward a theory of social practices: A development in culturalist theorizing. *European Journal of Social Theory, 5*(2), 243–263.

Ryan, M., & Blok, V. (2023). Stop re-inventing the wheel: Or how ELSA and RRI can align. *Journal of Responsible Innovation, 10*, 1–19.

Schatzki, T. R., Knorr-Cetina, K., Von Savigny, E., et al. (2001). *The practice turn in contemporary theory* (Vol. 44). Routledge.

Sharon, T. (2021). Blind-sided by privacy? Digital contact tracing, the Apple/Google API and big tech's newfound role as global health policy makers. *Ethics and Information Technology, 23*(Suppl 1), 45–57.

Shneiderman, B. (2020). Bridging the gap between ethics and practice: Guidelines for reliable, safe, and trustworthy human-centered AI systems. *ACM Transactions on Interactive Intelligent Systems (TiiS), 10*(4), 1–31.

Smuha, N. A. (2021). Beyond a human rights-based approach to AI governance: Promise, pitfalls, plea. *Philosophy & Technology, 34*(Suppl 1), 91–104.

Swierstra, T., Stemerding, D., & Boenink, M. (2009). Exploring techno-moral change: The case of the obesitypill. In *Evaluating new technologies: Methodological problems for the ethical assessment of technology developments* (pp. 119–138).

Taeihagh, A. (2021). Governance of artificial intelligence. *Policy and Society, 40*(2), 137–157.

Taylor, L., Sharma, G., Martin, A., & Jameson, S. (2020). *Data justice and Covid-19*. Meatspace Press.

Verbeek, P.-P., & Tijink, D. (2020). *Guidance ethics approach: An ethical dialogue about technology with perspective on actions*. ECP Platform for the Informatiesamenleving. https://ecp.nl/wp-content/uploads/2020/11/Guidance-ethics-approach.pdf

Chapter 7
The Nature of Algorithms and Their Relation to Accountability

Mario Verdicchio

Abstract This work carries out an analysis of algorithmic accountability from a computational perspective, that is, it tackles the question of how accountability relates to algorithms with a focus on the nature of the latter. The aim is to provide an analysis of what algorithms are and show that their abstract character sets them apart from the concrete problems of bias that have ignited many debates on accountabilty in recent years. This does not mean that algorithms are not part of these problems, but that what is achieved with algorithms is just one aspect of an articulated process that includes so much more, both within and outside the context of computer science. There is a fundamental distinction between an algorithm that is only described on paper and one that is actually executed. Moreover, whether the executor is a person or a computer and, in the latter case, whether the programming follows a top-down or a bottom-up approach can make a difference. All these variations around the execution of an algorithm are crucial when it comes to any attempt at defining accountability in this context, so any relevant discourse needs to take them into account.

Keywords Accountability · Algorithms · Computers · Machine learning · Programs · Software engineering

---

M. Verdicchio (✉)
Department of Foreign Languages, Literatures, and Cultures, University of Bergamo, Bergamo, Italy

Berlin Ethics Lab, Technische Universität Berlin, Berlin, Germany
e-mail: mario.verdicchio@unibg.it

M. Noorman, M. Verdicchio (eds.), *Computer Ethics Across Disciplines*, Philosophy of Engineering and Technology 44,
https://doi.org/10.1007/978-3-031-87974-6_7

Introduction

Algorithms are increasingly feeding into decisions about credit, insurance, recruitment, and content selection on social media and streaming services, and the problem of accountability in relation to algorithms has become important, since issues of bias have often emerged and raised public concern (Shah, 2018).

This chapter tackles the question of how accountability relates to algorithms from a computational perspective, that is, it does so with a focus on the nature of algorithms. The aim is to provide an analysis of what algorithms are and show that their abstract character sets them apart from the concrete problems of bias in race (e.g. soap dispensers unable to detect hands with dark skin, Fussell, 2017), gender (e.g. STEM job advertisements shown more often to men than women on social networks, Lambrecht & Tucker, 2019), or social status (e.g. loans granted or denied on the basis of applicants' affiliations, Hurley & Adebayo, 2016).

This does not mean that algorithms are not part of these problems at all, but that what is achieved with algorithms is just one aspect of an articulated process that includes so much more, both within and outside the context of computer science, that "algorithmic accountability" may even be considered a misnomer. Still, when it comes to computational endeavors, algorithms are the starting point of the discourse, so I will first focus on their very definition.

Definition of an Algorithm

One of the simplest and yet comprehensive definitions of algorithm is provided in an appendix to an article by philosopher W. J. Rapaport on the computational nature of human cognition (2012). Despite the very ambitious goal of that paper, which is part of the decade-long debate on the use of computation to either simulate or produce mental processes sparked by Alan Turing's pioneering work (1950), the definition is very down-to-earth and easily comprehensible also by those not specialized in computer science.[1] Here it is:

> An algorithm (for executor *E* to accomplish goal *G*) is a procedure, that is, a finite sequence (or set) of statements (or instructions) such that each statement is: comprised of a finite number of symbols from a finite alphabet, unambiguous for *E* (i.e., *E* knows how to do it, *E* can do it in a finite amount of time, and *E* knows what to do after its completion); moreover, the procedure takes a finite amount of time and when it ends, *G* has been accomplished.

Two aspects stand out immediately: relative to an algorithm, there is an executor *E* and a goal *G*. Interestingly, there is no specification on the nature of *E*: it could be a person *P* or a machine *M*. Of course, depending on who or what is the executor, the

[1] Hill (2016) provides an extensive list of references to various definitions of algorithm, also by computer scientists, and she too selects Rapaport's proposal for a discourse that is directed to philosophers of technology.

meaning of some parts of the definition of an algorithm changes: "*E* knows how to do it", for instance, has a familiar meaning when *E* is a person (i.e. we all know what it means to know or not know how to perform a task), whereas it calls for a series of activities like modelling, designing, and metaphorical thinking when *E* is a machine *M* and we state that "*M* knows how to do it."

The nature of executor *E* is important also for the relation between *E* and goal *G*. If *E* is a person *P*, they may entertain *G* as their own goal: *P* wants to achieve *G*, concocts a relevant algorithm, executes it, and in the end they succeed. If *P* does not have such goal, *P* may still be the executor of the algorithm on behalf of another person or group of persons, aiming at achieving the goal.

Accountability enters the picture when the achievement of the algorithm's goal produces results that are considered damaging, harmful, or unfair by somebody. When no machine is involved, and person *P* has executed the algorithm, some questions may rise when such execution has caused concern about whether *P* has designed the algorithm and whether they have executed it correctly. These questions are related to each other because so are the factors they are pointing at. These are factors intervening in the design and the use of algorithms, which are in complex relations that may be neglected until something goes wrong and the need for accountability sheds light on them. As pointed out by its definition, an algorithm is designed with a goal *G*. If *P* has indeed designed the algorithm, then *P* has done so with goal *G* in mind. In computer science, an algorithm is said to be correct with respect to a specification if its execution runs as prescribed by such specification. Hence, if *P* has designed an algorithm with goal *G*, the algorithm is correct if it achieves goal *G*. If harmful effects have called for accountability on the algorithm, another legitimate question is whether they were the goal of the algorithm, or part of it, or whether they are a side effect of the execution. A harmful goal on the one side and harmful side effects of an apparently harmless goal on the other make for very different cases of accountability. I will start here with the former, and deal with the latter later.

Here is an example algorithm, which is extremely simplified with respect to the reality of the context it refers to, i.e. evaluation of loan applications, but may help analyze the questions above and clarify the role of algorithms when it comes to accountability. The algorithm is presented with the graphical device of a flow chart, where each block has a different shape depending on the kind of instruction it contains, and all blocks are connected to each other with an arrow that represents the flow of the algorithm's execution.

Figure 7.1 depicts the different types of instructions an algorithm is comprised of: the circle from which an arrow departs marks the starting point of the algorithm; the parallelogram corresponds to an input instruction, which brings data from the outside world to the executor (in this case, the loan applicant's data); the rhombus represents a condition check, where the execution takes a specific path in accordance with the result of such check; the block shaped like a document sheet indicates an output instruction, which sends the result to the outside world (in this case, the result of the application); finally, all paths lead to the end of the algorithm, marked by another circle.

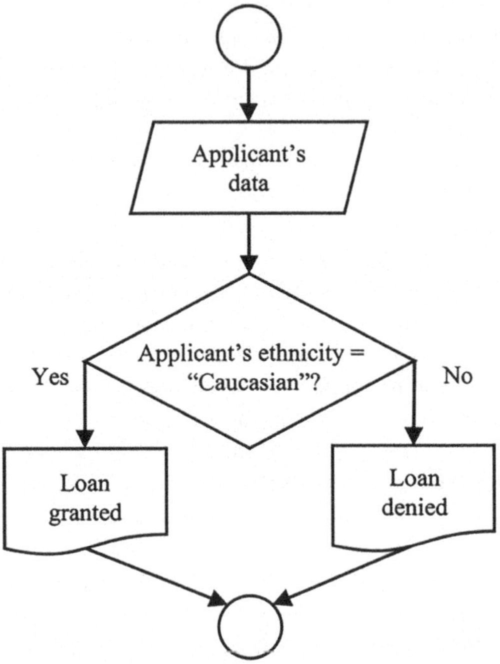

Fig. 7.1 Flow chart of an algorithm for the evaluation of loan applications

Obviously, the depicted algorithm contains an instruction that reflects a racist way of managing loan applications.[2] Imagine that the executor *P* is an employee of a bank *B* and that *P* processes loan applications made to bank *B* by following the algorithm in Fig. 7.1. Even in this example, made so simple to the point of unrealism, issues of accountability rise in a full-fledged form. If accountability, as defined in the seminal work by Bovens (2007), is a relationship between an actor and a forum towards which the actor must explain their acts with possible consequences, then a number of conditions must hold for accountability to be possible in the scenario above, and they all hinge on the fact that an algorithm gets executed. This may seem obvious: for an algorithm to have consequences it needs to be set in motion. Less obvious is the fact that the characteristics of a written algorithm and those of an executed one present a significant difference. The clarity with which all instructions must be specified inside an algorithm by definition (including those reflecting a racist policy) is not necessarily there when the algorithm is executed, since the execution of an algorithm does not entail any public declarative act by the executor. Indeed, the only points of contact between the algorithm's instructions and the

[2] I will try to refrain from writing expressions like "racist instruction" because attributing values, whether positive or negative, to a non-sentient item like a component of an algorithm is an example of those metaphors, anthropomorphisms, and projections that jumble the discourse on artificial entities and moral values, responsibility and, hence, accountability. See Verdicchio and Perin (2022) for a discussion on this issue.

outside world are given by the input instruction, with which the applicant's data are inserted, and the output instruction, which may result in a "Loan granted" or "Loan denied" outcome.

The concept of a "black box" has often been used in recent years to refer to specific types of Machine Learning algorithms whose execution is characterized by opacity (Carabantes, 2020), but even the simplest of algorithms, when executed by a person who does not make any declaration if not that prescribed by the output instruction, can be a black box. However, there is a difference. Despite being introduced in Machine Learning, the black box metaphor is not perfectly fitting in that context because even if we look inside a Machine Learning system that has provided a certain result, we are not able to find an explanation for such result: it is as if we opened the black box, and we stayed as clueless as before about its content. I will get back to the opacity and accountability issues in Machine Learning later. Instead, in the case of a very simple algorithm like the one in Fig. 7.1, the black box metaphor is appropriate, because opening it provides all the access and visibility that we need to explain the result of the execution of the algorithm. In this case, opening the box means obtaining access to the instructions that constitute the algorithm and uncovering the racist policy for the evaluation of loan applications in the form of a condition check.

There is a stark contrast between the blatant racism shown in Fig. 7.1 and the reticent behavior of a bank employee that simply states that a request of a loan has been rejected. There is no box to open when we are dealing with an algorithm because by its very definition everything must be clear and unambiguous. Borrowing another metaphor, we would state that the algorithm is an open book. Instead, when an algorithm is executed, either by a person or a machine, such clarity is lost, because the rest of the world, including the users who are possibly damaged by such execution, can only see the interface, i.e. the wrapping outer layer that the executor is presenting. This difference is crucial when it comes to accountability, so any discourse on accountability and algorithms needs to take it into account. The fundamental kick-off question is what it means to put an algorithm into execution.

From Algorithms to Programs

How an algorithm is executed largely depends on the nature of its executor *E*. To emphasize the fact that algorithms are not essentially related to computer science and computing machines, they have been executed by persons for the vast majority of recorded human history, starting with the Sumerian clay tablets on number division dating back to 2500 BCE (Chabert, 1999).

When person *P* executes an algorithm, their course of action must be limited to what indicated by the algorithm itself.[3] As stated before, if an algorithm is correct, it means that its execution will deterministically, inevitably lead to the achievement of the goal *G* for which it was conceived in the first place. With the standing assumption of correctness, person *P*'s task is to follow the algorithm step by step, without any deviation. Depending on how long and complex the algorithm is, *P* may need a memory support in the form of a written document that lists all the instructions, or they may learn the whole algorithm by heart and execute it.

The despicable algorithm in Fig. 7.1 is, indeed, very short and simple and, hence, easy to memorize. In a setting where *P* is a bank employee running that algorithm by heart, a number of conditions must hold for the issue of accountability to emerge in full, that is, for the racist policy to be brought to light. The following conditions are not independent: except for the first one, each condition depends on the occurrence of the previous one. The algorithm must be executed (1); the execution of the algorithm must trigger a reaction in the people affected by it (2); the reaction must lead to an analysis of the algorithm (3); finally, the analysis must determine the existence of a problem in the algorithm (4).

In a computer-less, exclusively people-based scenario, not all these conditions are easy to meet. If bank *B* not only has a racist policy but is also very secretive about it (precisely to avoid social stigma and legal complaints by the relevant forums of accountability), its management might have instructed executor *P* to divert attention away from the most critical part of their algorithm by means of prepared responses. An interaction with an inquiring non-Caucasian customer whose loan request has just been turned down might be as follows:

P: "Unfortunately, we cannot accommodate your request."
Customer: "Why?"
P: "Your financial profile does not meet our eligibility criteria."

At this point, conditions 1 and 2 have been met: *P* has executed the algorithm, and the result has triggered a reaction in the turned-down customer. However, condition 3 appears to be rather hard to realize, if only the resources at the customer's disposal are considered. The customer does not have access to the instructions of the algorithm that determined the decision. Since the algorithm has been learned by heart and run from memory by executor *P*, the customer has only the explanations fabricated by *P* at their disposal. The more realistic are the explanations (e.g., they are based on some weak point in the customer's financial history that has been designedly matched against a particularly strict criterion), the more likely is the success of *P* in convincing the customer that their evaluation was fair.

One way for the customer to detect that something is wrong is for them to observe that another customer, who happens to be identical to them in all financial aspects

---

[3] Of course, such limitation is in itself limited to the context of the algorithm's goal. For instance, a person is allowed to breathe while executing an algorithm even if it is not specified in the form of an instruction, since not only their breathing does not affect the algorithm's outcome, but it is literally vital for the executor.

but Caucasian, gets their request approved. However, it is unlikely that such a situation will arise, especially in consideration of bank secrecy rules and privacy surrounding people's finances. There should be a group of investigators disguised as customers asking for a loan with financial data designed to provide significant outputs from bank *B* by which they could perform racism-detecting inferences like the one described above.

This imagined investigation points to a concrete technique that is used in the context of Machine Learning to infer information about black-box algorithms. The basic idea is to build a model of an algorithm whose rules are not known by feeding it mock data points that are designed to differ from each other only in few key aspects. The differences in the corresponding outputs should provide cues on how the algorithm works and what rules are governing the decisions at least locally, that is, in the proximity of the data points used in the analysis (Ribeiro et al., 2016).

Whether mock or real, data are a fundamental component of algorithmic practice in general, not only in Machine Learning, and their presence marks a shift from theoretical reasoning regarding a plan of action to the actual execution of that plan. Indeed, what depicted in Fig. 7.1 is only a blueprint or a script for actions, but until those instructions are given to an executor, an algorithm does not entail any actual act. The racism is glaring, but still only potential. Correspondingly, the string "Applicant's data" does not refer to actual data but works as a placeholder that allows to describe now what to do with the data that will arrive in the future, if ever the algorithm is executed.

The chasm between an algorithm on paper and one in execution is even more evident when executor *E* is a machine *M*. One of the definitory characteristics of an algorithm is that its instructions are unambiguous to *E*. For a person *P* who is familiar with the context in which an algorithm is conceived and who is fluent in the language used to write the algorithm's instructions we can easily assume, also on the basis of our own experience as human beings, that all instructions are unambiguous. If *P* is also the creator of the algorithm, this is obvious; otherwise, *P* has the possibility to consult with the creators of the algorithm to eliminate any ambiguity before starting the execution.

If the executor is a machine *M*, all these considerations do not hold, because *M* lacks both familiarity with the algorithm's context and comprehension of the language, for the simple reason that *M* is not a person.[4] The algorithm in Fig. 7.1 is immediately comprehensible to a person who speaks English and is familiar with the visual semantics of flow charts but it cannot be given to a computer for execution. Computers are machines built in a way that prescribes a very specific technique for giving them instructions, which is not what one does with human executors, that is, showing them a figure with text printed on paper. Such technique consists in creating a sequence of commands inside a computer's memory that mirror the instructions of the algorithm.

[4] These limitations have been discussed in the long-standing debates on the relations between computers and common sense (Dreyfus, 1992), and between computers and semantics (Searle, 1980).

In computer science terms, if we want a machine *M* to execute the algorithm, we need to write the relevant program inside *M* or, in other words, we need to program *M* accordingly. Many computer scientists refer to this technique as a conversion of an algorithm into a program, but this is a misconception. An algorithm does not disappear because it has been transformed into a program in the way the ingredients of a recipe disappear when they become part of the cooked dish: the algorithm keeps on existing just like the recipe on the cookbook, to be used time and time again. As the faithfulness of the result in the kitchen to the perfect dish depicted in the cookbook depends on the dexterity of the cook, so the adherence of the written program to the algorithm depends, among many factors, on the talent of the person who wrote it, commonly known as the programmer.

The other determining factors that contribute to the creation of an apt program are associated with the design and construction of machine *M*. Indeed, when a programmer is to write a program, not only they assume that the algorithm upon which the program will be based is correct, but they also take for granted that the machine in which they are going to write the program works properly.

Computers, understood as computing machines, have taken different shapes through the centuries, but we can mark a significant technological boundary in the 1950s, when the discovery of semiconducting materials enabled the invention of transistors, which brought the era of mechanical computers to an end and established digital electronics as the new universal standard for automated computation (Garfinkel & Grunspan, 2018). The electronic circuitry of which computers are nowadays comprised is a technological embodiment of the fundamental principles that govern how data are elaborated in these machines: everything that is to be given to a computer, both instructions and data, must be binarily encoded into low-tension and high-tension electrical signals (interpreted by computer scientists as '0' and '1', respectively), which are conveyed into, stored in, transformed, and sent out by the circuits in the computer.

From this perspective, writing a program on a computer means inserting and storing a sequence of electrical signals that, once they are travelling through the computer's circuits, make the computer perform operations and yield results. The circuits that enable the input of signals (and also their output) constitute a computer's interface; the circuits that store the signals are a computer's memory; finally, the circuits that elaborate and transform the signals constitute a computer's processor. For a programmer to take for granted that the machine they are using works correctly means to assume that all these circuits have been designed and built in perfect accordance with their specifications. This has not always been the case, as I will show in the next section.

The Faults in Machines and in Programs

In June 1994, mathematics professor Thomas Nicely of Lynchburg College in Virginia noticed that after adding a new computer with a Pentium processor to the roster he was using to conduct experiments on prime numbers, he started obtaining results that were inconsistent with mathematical theory. Nicely spent months isolating various factors that could be causing these errors, but in the end, it became clear that they did not depend on an error in the program he had written. Instead, it was the processor of the newly added computer that was performing certain divisions incorrectly. Since it was one design flaw in a specific point of the electronic circuits, only divisions of particular sequences of digits involved the faulty part and therefore led to incorrect results. Intel, the company manufacturing the processor, stated that the average user would receive incorrect results from the processor once every 27,000 years, whereas according to the calculations by IBM, then a competitor of Intel in the processor market, that error would occur every 24 days of ordinary usage. In December 1994, under public pressure Intel recalled the defective processors, with an estimated corporate loss of around $475 million (equivalent to almost 1 billion dollars of today) (Cipra, 1995).

Nicely's case presents significant differences to the one of the non-Caucasian customer of bank *B*. Firstly, having a computing machine as executor sheds light on the necessity of referring to programs instead of algorithms, because what runs inside a computer is a concrete program with real data in the form of electrical signals and not an abstract course of future action written in natural language with placeholders instead of data.[5]

When a person *P* is the executor, we are not able to draw a clear distinction nor trace relationships between what described in the algorithm and the actions taken by *P*, since we do not have access to the inner workings of a human brain, in particular those that lead from the perception and comprehension of written instructions to the relevant enactment in terms of thoughts, language and bodily movements.

On the contrary, when the executor is a machine *M*, there is a very clear, ontological distinction between the instructions written in an algorithm and those stored in a *M*'s memory, because the format of the latter is completely determined by the design of the machine: the architecture of the processor determines what kind of operations the machine is able to perform, and the task for the programmer is to map

[5] To adhere strictly to computer science, what runs inside a computer should be called a process. A program is the set of instructions stored in the memory, more static and distinct from the dynamics of what happens in the processor. To make this distinction clear, imagine the word processor installed in your computer: the installed program (e.g. Microsoft Word) is one, and resides in the memory, whereas you can launch the program several times to work on different documents at the same time, thus creating as many processes. For this discourse on accountability, the distinction between programs and processes is not particularly meaningful (I would have to refer to extreme cases of real-time computing in which different processes compete for computing resources) and will be omitted.

the instructions in the algorithm onto appropriate combinations of those machinic operations.

A correct mapping is fundamental for the goal of the algorithm to be reached by means of the execution of the relevant program on a machine, and this is another point where Nicely's case stands out. By adapting the accountability conditions from the context of an algorithm executed by a person to the one of a program executed by a machine, we have the following: the program was executed (1), the execution of the program triggered a reaction (2), and the reaction led to an analysis of the program (3). Nicely's analysis led to groundbreaking results: condition 4, which prescribes that the analysis must determine the existence of a problem in the program, was not met because the problem was not in the program, but in the machine. Even worse, it was not a faulty instance of a machine: there was a design flaw in the blueprint of the machine, which meant that a great number of faulty machines had already been manufactured and sold. Nicely was able to rule out any problem in his program because of the special nature of the context in which he worked: number theory. This means that the professor was able to map the operations of his program back onto the passages of the theorems he was using to work with prime numbers. In other words, he was relying on a machine only to execute more efficiently something that could be carried out by a mathematician with algorithms whose correctness was guaranteed by theorems verified within the discipline. If a provenly correct theory yields a certain result and the relevant program does not, it is very reasonable for the programmer to doubt their own mapping process, but in Nicely's case, the mapping was guided by the theory itself, hence he could divert his analysis to the only other possible although unlikely culprit: the machine. This incident is unique in the history of mass-produced computing machines, but it shows that switching from persons to machines for the execution of algorithms adds a number of possible points of failure and hence calls for accountability, including the constructors of the very machines we use to automate computation.

Most programmers do not work with the support of mathematical theorems, because most contexts in which algorithms are conceived and programs are written have little to do with mathematics or any other kind of formal theory. Indeed, the nature of a programmer's activity, also known as software development, is dubbed as informal by the programmers themselves (Hailpern & Santhanam, 2002). Without a consolidated theory that provides rigorous proofs on the correctness of algorithms and their mapping onto programs, software developers are left to their own devices and, if a faulty design in a circuit is an extremely rare event, errors in algorithms and programs are much more frequent.

The cause of such errors is obvious: algorithms and programs are written by humans, and humans are fallible. An adage by the author of "Murphy's law" (Bloch, 2003) captures the nature of this fallibility well: "A computer program does what you tell it to do, not what you want it to do." Indeed, a programmer's mission is to bridge the gap between a goal to be reached, often expressed in vague, abstract terms, and the very precise and concrete set of instructions to be executed: the wider is the gap, the more difficult is the mission and, hence, the more likely it is for some mistakes to occur. Such mistakes can be divided in two categories that mirror the traditional distinction between syntax and semantics.

Firstly, a programmer might write instructions that are syntactically defective, that is, they do not follow the rules imposed by the machine the instructions are meant for: for instance, to make the computer print a result, the programmer might write "rpint" instead of "print" in the relevant instruction.[6] Syntax errors are easy to detect because their identification is equivalent to the task of determining whether a given string of characters belongs to a specified language or not, and this is a problem that has been successfully investigated in the field of theoretical computer science since its beginnings (Hopcroft & Ullman, 1969), and it is now implemented in all computer systems. Thus, while they are writing a program on a computer, programmers are usually assisted by another, already complete and running program, a compiler, that checks for syntax errors.

Secondly, and much more dangerously, a programmer might write an instruction that is syntactically correct, but whose effects not only do not constitute a step towards the goal for which the program is being written, but they might even work against it. For instance, an algorithm prescribes that the sum of value a and value b be assigned to v, which is an operation whose relevant instruction may be written as "$v \leftarrow a + b$"; suppose that, because of a trivial typing error, the programmer writes "$v \leftarrow a - b$" instead. The written instruction is taken by the computer at face value: assign the difference between a and b to v. This is a semantic error, where semantics is meant in the restricted, operational sense adopted in computer science: the meaning of an instruction is what that instruction does in terms of state change in the computing system where it is performed. In this sense, semantics is machinic behavior (Goguen, 1975).

Semantic errors are more dangerous because, since there is no problem with the syntax, they cannot be detected by a compiler while the program is being written, a phase that is known in software engineering as "compile time". Instead, they can reveal themselves only at "run time", that is, when the program is executed, and all accountability conditions hold: since the results do not match the programmer's expectations, the execution of the program triggers their reaction, leading to a reexamining of the program which, in turn, allows them to find the error.

Differently from the syntax rules of an artificial language, a programmer's goal and intentions cannot be encoded into a computer; thus, there is no automated tool that can quickly detect semantic errors. A subfield of software engineering called "software testing" has been developed through the years on the basis of the practical experience and knowledge of programmers, who compare their activity to an art, so far it is from being fully captured in the form of protocols, to-do lists, and

---

[6] Programs are binary electrical signals inside a computer's memory, but this does not mean that programmers must produce binary electrical signals. So-called programming languages exist, which enable programmers to write instructions in an artificial language that is designed to be much closer to English. For each programming language (e.g. Python, Java, C, etc.) there is a "compiler", which is a software that automatically translates the instructions written in that language into binary electrical signals inside a computer memory.

algorithms (Myers et al., 2004). Before being deployed in the real world, a program must be tested as much as possible in the computer lab where it was written. In its simplest form, testing means feeding the program with as many mock data as possible, and matching the obtained results with the programmers' desiderata. An exhaustive exploration is not possible. From a theoretical perspective, since a computer works with numerical data, software testers should feed it with all possible numbers, which would delay the deployment indefinitely. From a practical perspective, even the most powerful computer has a finite amount of memory, which puts a cap on the biggest number it can store and elaborate; still, a complete test is unfeasible because trying every possible value on each piece of data involved in the program would mean exploring all possible combinations of those numbers, which would lead to a combinatorial stacking up of tasks and, again, a delay that is incompatible with the whole endeavor.

Software testers have the mission of selecting a set of testing data that is a compromise between the need for a thorough check of the program and the need for a timely deployment of it. Experience and common sense guide this process: for instance, it does not make much sense to test the effects of numerical inputs much greater than 100, when the relevant instruction processes an applicant's age in the context of a loan request.[7] Once the testing phase is considered complete by the testers, the program is deployed, sometimes with disastrous effects.

One of the most sensational cases of faulty software in recent years happened in the United Kingdom over more than a decade since the first informal investigations started in 2009 thanks to an appeal from Lee Castleton to the IT magazine "Computer Weekly" (Thompson, 2009). Castleton is a former postmaster (head of a post office) who faced bankruptcy after refusing to pay the Post Office £27,000, claiming that he did not owe the sum despite the computerized accounting system, called Horizon and developed by software company Fujitsu, showed a deficit, for which postmasters are personally responsible in the UK. Castleton contested the debt in court, but he lost, which led to bankruptcy and £321,000 in legal costs. Computer Weekly had already been contacted in 2004 by another postmaster, Alan Bates, but back then it was wrongfully considered an isolated, non-significant case. The investigations by Computer Weekly revealed very similar disputes with at least six other postmasters, who all cited issues with the Horizon system and a lack of support from their union. Not only the Post Office denied any software-related faults and complaints from postmasters, but they even resorted to sending individuals to prison as a deterrent: Seema Misra, another former postmaster, was prosecuted and imprisoned based on

[7] Sometimes, numerical workarounds in software come back to bite in unexpected ways. Many systems developed in the twentieth century were using only two figures "YY" to store the year in chronological data, which caused the Y2K panic towards the end of 1999, when many thought that the ambiguous value of "00" (1900 or 2000?) would cause severe malfunctions in computer-controlled systems. Nothing noteworthy happened on January 1st, 2000, but significant disruptions occurred in many cash dispensers around the world on February 29th, 2000, since 1900 was not a leap year whereas 2000 was, so that these systems went from "02/28/00" to "03/01/00", skipping the actual date of "02/29/00" (Wired, 2000).

evidence provided by the Horizon system. Shortly after Computer Weekly highlighted the challenges faced by some postmasters, several others who had encountered similar issues gathered and sparked a nationwide campaign. When echoes of such campaign reached the Parliament and some of its members raised concerns, the government-owned Post Office faced pressure to address the issues and in 2012 it hired Second Sight, a forensic accountancy firm, to initiate an external review and mediation program. Many suspected that the review was a façade for a cover-up; instead, Second Sight discovered that postmasters were not responsible for the cash discrepancies but were struggling with the flawed Horizon system and the Post Office's prioritization of its reputation over the welfare of employees. After an interim report in 2013 that raised serious concerns, just before the full investigation report's release in 2015, Second Sight's work was halted, and the mediation scheme was terminated, which convinced the postmasters' campaign organization to proceed with a class action, with hundreds of victims seeking damages from the Post Office for the harm caused by the unreliable Horizon system, despite the Post Office's continued assertion of its correctness. In December 2019, after two trials, one examining the Post Office's contract with the postmasters, and the other focused on the Horizon system, the Post Office settled with the postmasters after initially trying to disqualify the presiding judge for bias. The 555 claimants received £58 million in compensation but, after expenses, only £11 million were left to be divided among them.

The case confirmed the postmasters' claims about the flawed Horizon system, exposing software errors that led to accounting discrepancies. This revelation provided evidence for hundreds of previously convicted postmasters to appeal their convictions, leading to 93 overturned convictions out of over 736 cases related to Horizon. The government plans to introduce emergency legislation to address the remaining convictions as a group. Prime Minister Sunak labeled this case as "one of the greatest miscarriages of justice in our nation's history," yet no Post Office nor Fujitsu executives have faced consequences. Instead, Fujitsu continues to secure government IT contracts, and Paula Vennells, CEO of the Post Office from 2012 to 2019, left the organization with a large payout. The only individuals penalized so far have been the victims of this scandal (Flinders, 2024).

The Post Office debacle brings the accountability conditions to a whole other level, where the execution of a faulty program (1) does trigger a reaction in the people affected by it (2), in terms of wrongful incarcerations and campaign organizations, whose attempts to ensure an analysis of the program are actively opposed by corporate and political power; when such analysis is finally granted in court (3), and it does determine the existence of a problem in the program (4), none of those responsible for the errors in the program nor those responsible for the attempted concealment of evidence are held accountable. This situation is just as deplorable as the dystopian fantasy of racist bank B, yet regrettably, it is also real.

This case sheds light once more on the crucial role played by data. One isolated case was not considered worth further inquiry by Computer Weekly in 2004: a second, very similar instance 5 years later triggered the investigation. This is an example of a very general way of thinking that seems hard to dispute: having adequate evidence for a hypothesis gives us reason to believe that such hypothesis might be true, and the more is the evidence the more justified are our beliefs. If data can serve as evidence, then big quantities of data may mean that we have more chances to confirm some of our hypotheses or even discover new knowledge. This perspective is at the foundation of the synergy between the fields of Big Data and of Knowledge Management (Erickson & Rothberg, 2015) which, in connection with more powerful processing power in computers and increased data exchange over the Internet, has enabled the transformation of Machine Learning from a theoretical research on algorithms studied in journals (Leung & Haykin, 1991) to a programming paradigm in the software industry and beyond (Messaoud et al., 2020). I will explore the impact of the use of data on issues of accountability in what follows.

The Growing Role of Data

Machine Learning is a subfield of Artificial Intelligence that has gained worldwide traction in the past decade thanks to notable successes in fields like the game of Go (Silver et al., 2018) or essay composition (Biever, 2023), which were traditionally considered exclusive to humans, given the considerable role played by intuition and creativity in those tasks. Making computers do something that is typically human is the founding characteristic of Artificial Intelligence in general and, thus, of Machine Learning in particular. However, Machine Learning is based on a paradigm that is reversed with respect to what proposed by the Artificial Intelligence of the early years.

The main assumption shared in the 1950s by the pioneers of the field was "that every aspect of learning or any other feature of intelligence can in principle be so precisely described that a machine can be made to simulate it" (McCarthy et al., 2006, p. 12). This approach is in line with what illustrated so far about algorithms and programs: there is a goal G we want to achieve; we analyze the context of such goal in terms of means and constraints and conceive an algorithm; if that algorithm is to be given to a machine M for execution then a programmer maps the algorithm onto the operations of M and creates a program; finally, M executes the program and G is achieved. This is how programmers have operated traditionally: from the abstract concepts of algorithms to the concrete operations of programs, from the formal rules with placeholders to the practical operations on real data. If we accept a conventional hierarchy in which abstract entities are above and concrete ones are below, then this way of creating programs is a top-down approach. Professor Nicely's programming activity was top-down: starting from the axioms of number theory, all the way down to division of numbers in a (faulty) processor. The Fujitsu software followed the same paradigm: the company started from the high-level specifications of accounting goals of the Post Office and worked its way to the now infamous Horizon system that was meant to achieve them. Without the support of a rigorous theory, the latter endeavor ended up including several human errors.[8]

Machine Learning, instead, is based on a bottom-up approach, where the starting point is a set of concrete data from the real world that are analyzed to allow for the inference of a general mathematical model of which those data can be considered a particular instance. The paradigm is the opposite of the previous one because it goes from specific, concrete cases to a general, abstract theory. This is where the abundance of data made available by the Internet plays a key role: the more data points the Machine Learning system has, the more precise the mathematical model can be. The metaphor underlying the name of this approach is thus clear: by means of data analysis, a computing machine can "learn" the general rule governing the values of those data.

Getting back to the example of bank *B*, suppose that the bank's management is compelled by a federal court to comply with the Equal Credit Opportunity Act, which makes it illegal for banks to consider certain factors, including race, when evaluating a loan application. This means that the despicable algorithm depicted in Fig. 7.1 and, more precisely, the condition specified in the condition check must not be used anymore, and the bank needs to come up with a new, law-abiding one. The rest of the algorithm can be left as it is, since it only consists of an input instruction to obtain the applicant's data and an output instruction to issue the result of the evaluation (Fig. 7.2, left).

For bank *B* to opt for a bottom-up, Machine-Learning-based approach to the formulation of the condition, there is the need for real data from past loan application evaluations to analyze them from a mathematical perspective and build a model that is coherent with their values. Figure 7.2 (right) depicts a simplified visualization of this process.

First of all, since the model is supposed to be a mathematical function, all data must be provided in numerical form: this is easy for many aspects of an applicant, which are already numerical (e.g. credit score, savings, yearly income, etc.), whereas others need to be encoded, that is, mapped onto numbers with an arbitrary convention (e.g. residential address, educational level, etc.). The fully encoded, numerical data are then inserted in a Cartesian coordinate space, whose dimensions correspond to the various aspects of the applicants that were evaluated in the past. The various data points are not only characterized by their coordinates, but also by the result of the relevant evaluation, also encoded: since in this case the possible results are only two ("yes" and "no"), they will be typically encoded inside a computer as 0 and 1, respectively. In Fig. 7.2 the results are color-coded: the data points corresponding to an applicant who obtained the loan are black circles, whereas the others

[8] One might wonder whether Professor Nicely's program or the Horizon software, together with all the software written with a top-down approach are to be considered entities in the domain of traditional Artificial Intelligence. Indeed, since all kinds of software (on prime numbers, on post office accounting, on the game of Go, on essay composition, etc.) run on the same type of computing machine, there are legitimate reasons to question a rigorous partitioning of Computer Science into subfields (Verdicchio, 2023).

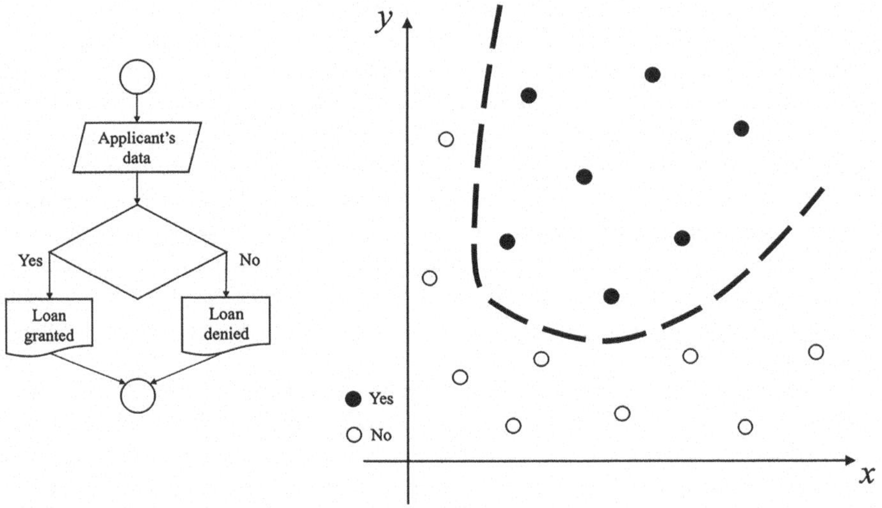

Fig. 7.2 A bottom-up approach to establish the condition in the algorithm for evaluation of loan applications

are white. Here only two dimensions (x and y) are depicted for visual clarity, so that the construction of the model boils down to tracing a curved line on a plane, but in real cases the model is a hypersurface inside a multidimensional space. Whether a line or a hypersurface, the desired mathematical model enjoys this property: it separates the space in two parts, one containing all and only the data points corresponding to successful applications, and the other with the data points of applicants who were turned down.

The dashed line in Fig. 7.2 is such model, which is "learned" by a particular kind of program that is called a "neural network." Its name is based on the original mathematical modelling that kickstarted in the 1940s the whole Machine Learning endeavor, even before digital electronic computers existed, when it was proposed to model a neuron, which either shoots an electric impulse or not, with a simple binary function, whose output is either 0 or 1 (McCulloch & Pitts, 1943). A neural network is a complex mathematical function comprised of a great number of extremely simple modules, called neurons, connected to each other in a series of layers, so that all neurons of each layer receive the outputs of the neurons of the previous layer and react to them producing an output that goes to the neurons of the successive layer (see Fig. 7.3).

A neural network "learns" by trial and error under the supervision of its programmers. The existing real data points are fed one by one to the network, whose structure (number of neurons per layer, number of layers) is one among many design factors that are decided on the basis of the programmers' empirical expertise. The bits (0's and 1's) of the binarily encoded input are distributed as inputs to the neurons of the first layer (Fig. 7.3 below left). Each neuron reacts according to its parameters (Fig. 7.3): the signal arriving along connection n is multiplied by a

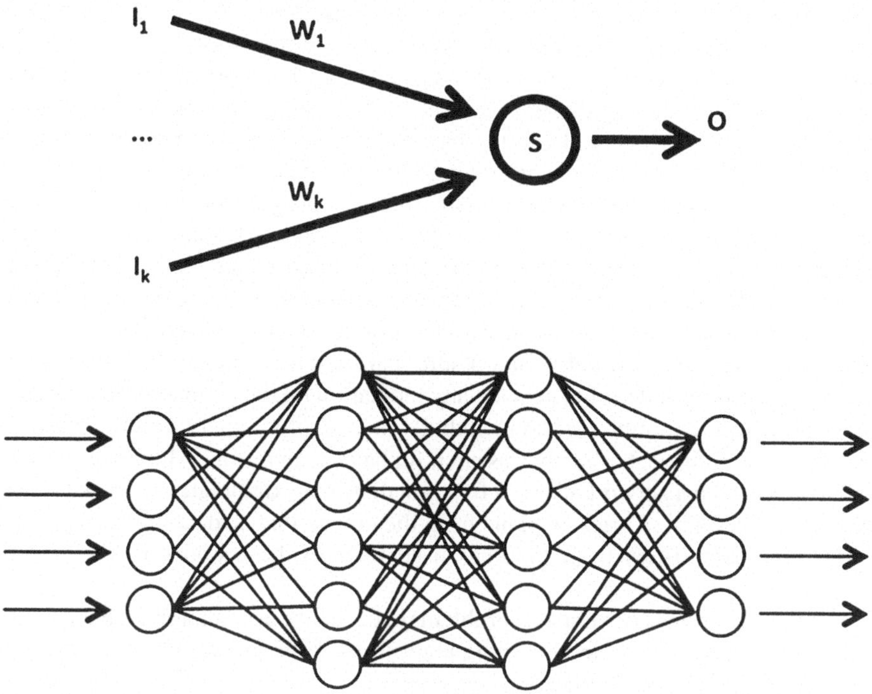

Fig. 7.3 An artificial neuron (above) and a neural network (below)

factor associated to that connection, called weight (w_n), and is summed to all the other input signals; if the total is greater than the neuron's threshold *s*, then the neuron "shoots" an output signal *o*, otherwise it stays dormant. Signal *o* works as an input to the neurons of the next layer the shooting neuron is connected to, and so on. All the output signals from the neurons in the last layer are combined to form the output of the neural network (Fig. 7.3 below right). What has been described so far constitutes a trial. The programmers feed an input to the network, i.e. data from a past loan applicant, knowing what result to expect. Suppose that the application was successful. If the bits in the output of the neural network are an encoded version of "no", then there is an error: the network has not computed the correct result corresponding to the given input and needs to be adjusted. This adjustment modifies the weights in the network to minimize the difference between the actual output and the desired one. The calculation of the adjustments is more efficient if it starts at the last layer and propagates backward through the network to the first layer, which is the why this technique is called "backpropagation." When the trial run is successful, that is, the network's output coincides with the desired one, no adjustment is necessary. This process is called training.

If the training goes well, from a certain trial on the network does not make errors anymore, and every input yields the output that the programmers are expecting on the basis of the past data at their disposal. This means that the neural network reacts to the data points exactly as depicted in the diagram in Fig. 7.2: it classifies applicant's data just as the bank employees who produced those data have done in the past. It is as if the neural network has "learned" the shape of the dashed line that separates the "yes" cases from the "no" cases, and it is ready to be used to classify new data. The new, Machine-Learning-based algorithm contains then the following condition check: "does the network reply to the applicant's data with a yes?"

There are a number of caveats. Firstly, there is no guarantee that the training comes to a successful conclusion. The mathematical theory behind neural networks warrants that they are universal approximators, that is, for any sort of classifying function, i.e. the dashed line in Fig. 7.2 that separates the data into categories, there exists a neural network that can learn it to any level of precision (Hornik et al., 1989). However, the theory does not assure that such network can be found in reasonable time, since it does not provide any indication on the number of neurons and layers needed: programmers need to embark on an exploration of different network structures and train each of them in search for satisfying results. Moreover, the precision with which a network learns a function does not extrapolate: a new data point that is very far on the Cartesian plane from the others, that is, data that present very different characteristics from the ones used to train the network is likely to be misclassified.

Misclassification is obviously a problem that calls for accountability, but here the black-box nature of Machine Learning algorithms is the biggest obstacle. Imagine the following exchange in the new bank *B*:

P: "Unfortunately, we cannot accommodate your request."
Customer: "Why? I thought I had everything needed to grant me a loan."
P: "Our neural network classified you as ineligible."
Customer: "Really? What of my credit history made it decide so?"

Even with the most transparent customer care and the willingness to explore the computational meanders of the machine that turned down their customer, employee *P* would only find a neural network, that is, a matrix of numerical parameters that correspond to all the weights that have contributed to the calculation of the result. None of them carry an explicit correspondence to the various components of the customer's data. The complexity of the internal organization of a neural network makes its inner workings uninterpretable by humans, including the very programmers who trained the network.

Luckily, there is now legislation against such a scenario: the abovementioned Equal Credit Opportunity Act in the United States and the General Data Protection Regulation in the European Union both include norms that safeguard the right to an explanation (Kaminski, 2021). Indeed, a new subfield of Artificial Intelligence, called Explainable Artificial Intelligence (Samek & Müller, 2019) has recently emerged with the goal of elaborating automated processes that shed light on the parts of the input that contributed the most to the final output of a Machine Learning algorithm and to support the formulation of an explanation.

In some cases, the explanation of an unexpected output has nothing to do with the input, but with the training data. In 2015, Google enhanced its photo management app with an automatic classification system aimed at labeling photos on a smartphone with keywords based on their content. In the United States, a black couple was labeled as "gorillas," marking the first case of a deplorable racist insult created by a Machine Learning system. It is suspected that the training of the neural network for person recognition was done only with Caucasian individuals (BBC, 2015).

This case circles back to the role that data play in Machine Learning systems: they counter a traditional top-down approach where programmers specify the evaluations criteria explicitly in the form of conditions with a bottom-up approach that builds a general evaluation function based on a myriad of data points. This means that, by construction, the data used to train a neural network also shape its fundamental characteristics. This also means that the quality of data determines the quality of the system, for the good but also for the bad (Chen, 2023). Imagine if bank B trained their new loan application evaluation system on the basis of their past racist evaluations. Data can work as evidence, it has been said, but if it is evidence of bias, then a Machine Learning system built on top of it could echo such bias in a computational form that is detectable only at run time, that is, only when the damage has already been done.

Conclusion

Algorithmic accountability pertains to determining the responsibilities for the outcomes of real-world actions resulting from decisions made based on algorithms. By illustrating the basic concepts and mechanisms in computer science that regard algorithms, this work was meant to show that an ascription of responsibility is not at all straightforward, because of the complex network of people and processes involved.

This is a non-exhaustive list of possible incidents that call for accountability:

- An algorithm may be not correct.
- An algorithm may be correct but based on a harmful goal.
- A person may execute an algorithm not correctly.
- A programmer might fail at writing a program correctly on the basis of an algorithm.
- A computer may be faulty in its processor, memory, or interface.
- There may be not enough time to test a program properly.
- An organization may trust their computers more than their employees.
- An organization may want to hide the fault in their software or computers for economic or political interests.
- A programmer may not be able to find a properly structured neural network to train.
- There may be not enough data or time for training a neural network well.
- A programmer may not be able to explain a wrong output of a neural network.

- A wrong output of a neural network may never be detected.

To call the relevant accountability "algorithmic" is limiting, because despite constituting a fundamental part of everything computational, algorithms alone do not do much. To go from a formalized idea of a course of action to the actual acts (and their consequences), there is the need for a complex apparatus that does include computing machines but, even more importantly, persons and organizations whose decisions fundamentally contribute to the achievement of the goals that were only vaguely envisioned when the relevant algorithms were conceived.

As Deborah Johnson and I have advocated for, it is important to avoid being "sociotechnically blind" (Johnson & Verdicchio, 2017) because focusing on the machines and overlooking the humans involved with these systems may mean missing a lot in terms of accountability, on both sides of the coin: not only the people and organizations that should be held accountable for the damages that they cause with the involvement of computational artifacts, but also the people and organizations that may play a decisive role in keeping the former accountable.

Algorithmic accountability starts with algorithms but the execution of such algorithms, which is what may ultimately cause harm, relies on many other factors, some of them (e.g. programs) often only informally based on algorithms, others (e.g. corporate decisions) not at all algorithmic, and heavily influenced by economy and politics instead. Accountability in algorithmic systems is complex and extends beyond just the algorithms themselves. This analysis is meant to emphasize the importance of considering the broader sociotechnical context in which these systems operate when striving for accountability. All efforts to ensure accountability should originate from an understanding of this context.

References

BBC. (2015). Google apologises for Photos app's racist blunder. *BBC News Tech*. https://www.bbc.com/news/technology-33347866 (Last visited in January 2024).

Biever, C. (2023). ChatGPT broke the Turing test – The race is on for new ways to assess AI. *Nature, 619*(7971), 686–689.

Bloch, A. (2003). *Murphy's law*. Penguin.

Bovens, M. (2007). Analysing and assessing accountability: A conceptual framework. *European Law Journal, 13*(4), 447–468.

Carabantes, M. (2020). Black-box artificial intelligence: An epistemological and critical analysis. *AI & Society, 35*(2), 309–317.

Chabert, J. L. (Ed.). (1999). *A history of algorithms: From the pebble to the microchip*. Springer.

Chen, Z. (2023). Ethics and discrimination in artificial intelligence-enabled recruitment practices. *Humanities and Social Sciences Communications, 10*(1), 1–12.

Cipra, B. (1995). How number theory got the best of the Pentium chip. *Science, 267*(5195), 175–175.

Dreyfus, H. L. (1992). *What computers still can't do: A critique of artificial reason*. MIT press.

Erickson, S., & Rothberg, H. (2015). Big data and knowledge management: Establishing a conceptual foundation. *Leading Issues in Knowledge Management, 2*, 204.

Flinders, K. (2024). Post office horizon scandal explained: Everything you need to know. *Computer Weekly*, January 3, 2024. https://www.computerweekly.com/feature/Post-Office-Horizon-scandal-explained-everything-you-need-to-know (Last visited in January 2024)

Fussell, S. (2017). Why can't this soap dispenser identify dark skin? *Gizmondo*, August 17, 2017. https://gizmodo.com/why-cant-this-soap-dispenser-identify-dark-skin-1797931773 (Last visited in January 2024).

Garfinkel, S. L., & Grunspan, R. H. (2018). *The computer book: From the abacus to artificial intelligence, 250 milestones in the history of computer science*. Sterling.

Goguen, J. A. (1975). Semantics of computation. In E. G. Manes (Ed.), *Category theory applied to computation and control* (Lecture notes in computer science) (Vol. 25). Springer.

Hailpern, B., & Santhanam, P. (2002). Software debugging, testing, and verification. *IBM Systems Journal, 41*(1), 4–12.

Hill, R. K. (2016). What an algorithm is. *Philosophy of Technology, 29*, 35–59.

Hopcroft, J. E., & Ullman, J. D. (1969). *Formal languages and their relation to automata*. Addison-Wesley Longman Publishing.

Hornik, K., Stinchcombe, M., & White, H. (1989). Multilayer feedforward networks are universal approximators. *Neural Networks, 2*(5), 359–366.

Hurley, M., & Adebayo, J. (2016). Credit scoring in the era of big data. *Yale Journal of Law and Technology, 18*, 148–216.

Johnson, D. G., & Verdicchio, M. (2017). Reframing AI discourse. *Minds and Machines, 27*(4), 575–590.

Kaminski, M. E. (2021). The right to explanation, explained. In S. K. Sandeen, C. Rademacher, & A. Ohly (Eds.), *Research handbook on information law and governance* (pp. 278–299). Edward Elgar Publishing.

Lambrecht, A., & Tucker, C. (2019). Algorithmic bias? An empirical study of apparent gender-based discrimination in the display of STEM career ads. *Management Science, 65*(7), 2966–2981.

Leung, H., & Haykin, S. (1991). The complex backpropagation algorithm. *IEEE Transactions on Signal Processing, 39*(9), 2101–2104.

McCarthy, J., Minsky, M. L., Rochester, N., & Shannon, C. E. (2006). A proposal for the Dartmouth summer research project on artificial intelligence, August 31, 1955. *AI Magazine, 27*(4), 12–14.

McCulloch, W. S., & Pitts, W. (1943). A logical calculus of the ideas immanent in nervous activity. *The Bulletin of Mathematical Biophysics, 5*, 115–133.

Messaoud, S., Bradai, A., Bukhari, S. H. R., Quang, P. T. A., Ahmed, O. B., & Atri, M. (2020). A survey on machine learning in Internet of Things: Algorithms, strategies, and applications. *Internet of Things, 12*, 100314.

Myers, G. J., Badgett, T., Thomas, T. M., & Sandler, C. (Eds.). (2004). *The art of software testing*. Wiley.

Rapaport, W. J. (2012). Semiotic systems, computers, and the mind: How cognition could be computing. *International Journal of Signs and Semiotic Systems (IJSSS), 2*(1), 32–71.

Ribeiro, M. T., Singh, S., & Guestrin, C. (2016). "Why should I trust you?" Explaining the predictions of any classifier. In *Proceedings of the 22nd ACM SIGKDD International Conference on Knowledge Discovery and Data Mining* (pp. 1135–1144).

Samek, W., & Müller, K. R. (2019). Towards explainable artificial intelligence. In W. Samek, G. Montavon, A. Vedaldi, L. K. Hansen, & K. R. Müller (Eds.), *Explainable AI: Interpreting, explaining and visualizing deep learning* (pp. 5–22). Springer.

Searle, J. R. (1980). Minds, brains, and programs. *Behavioral and Brain Sciences, 3*(3), 417–424.

Shah, H. (2018). Algorithmic accountability. *Philosophical Transactions of the Royal Society A: Mathematical, Physical and Engineering Sciences, 376*, 220170362.

Silver, D., Hubert, T., Schrittwieser, J., Antonoglou, I., Lai, M., Guez, A., Lanctot, M., Sifre, L., Kumaran, D., Graepel, T., & Lillicrap, T. (2018). A general reinforcement learning algorithm that masters chess, shogi, and go through self-play. *Science, 362*(6419), 1140–1144.

Thompson, R. (2009). Bankruptcy, prosecution and disrupted livelihoods - Postmasters tell their story. *Computer Weekly*, 11 May 2009. https://www.computerweekly.com/news/2240089230/Bankruptcy-prosecution-and-disrupted-livelihoods-Postmasters-tell-their-story (Last visited January 2024).

Turing, A. (1950). Computing machinery and intelligence. *Mind, LIX*(236), 433–460.

Verdicchio, M. (2023). Marking the lines of artificial intelligence. In S. Lindgren (Ed.), *Handbook of critical studies of artificial intelligence* (pp. 245–253). Edward Elgar Publishing.

Verdicchio, M., & Perin, A. (2022). When doctors and AI interact: On human responsibility for artificial risks. *Philosophy & Technology, 35*(1), 11.

Wired. (2000). Leap day had its glitches. *Wired*, March 1, 2000. https://www.wired.com/2000/03/leap-day-had-its-glitches/ (Last visited in January 2024).

Chapter 8
Prospective Algorithmic Accountability and the Role of the News Media

Nicholas Diakopoulos

Abstract Prior work on algorithmic accountability has focused on a retrospective understanding of the concept. In this chapter I suggest prospective accountability as a productive complement for addressing many of the underlying challenges to algorithmic accountability, including sociotechnical complexity and the various information deficits that stand in the way of retrospective blame assignment. I then elaborate a role for the news media to support prospective algorithmic accountability by contributing to norms for answerability and forward-looking responsibility assignments, as well as by helping to identify and normatively establish what is considered harmful and impactful in society and which therefore warrants mitigation. I argue that certain media activities such as editorial writing, solutions journalism, and predictive journalism could further enhance the media's impact towards supporting prospective accountability.

Keywords Prospective accountability · News media · Algorithmic accountability

Introduction

As societies around the world grapple with how to ensure that harms from algorithmic decision making systems and AI systems are mitigated, algorithmic accountability is a critical topic in today's technological landscape. However, much of the discourse on algorithmic accountability has been focused on conceptions of accountability as a retrospective activity of blame identification and assignment. Less attention has been paid to elaborating how to support prospective algorithmic accountability, which entails assigning forward-looking responsibilities to stakeholders in order to take particular actions and ultimately prevent undesirable

N. Diakopoulos (✉)
Northwestern University, Evanston, IL, USA
e-mail: nad@northwestern.edu

M. Noorman, M. Verdicchio (eds.), *Computer Ethics Across Disciplines*, Philosophy of Engineering and Technology 44,
https://doi.org/10.1007/978-3-031-87974-6_8

outcomes. The need for prospective accountability is motivated from several challenges facing blame assignment, namely that complex algorithmic systems obfuscate causality and information deficits frustrate the investigation and assignment of blame. A further difficulty relates to the need to establish the expectations for appropriate behavior within complex algorithmic systems. After laying out the background definitions of algorithms and accountability, and then problematizing the situation in terms of these various challenges, this chapter elaborates a role for news media.

In particular, the role for news media that I develop is largely grounded in the need to develop normative expectations for the human actors in algorithmic systems. This includes norms for answerability, norms for establishing what is harmful and impactful (i.e. outcomes or states of affair) that society may want to avoid, and then norms for assigning responsibility to particular actors for mitigating those potential states of affair. I suggest that the news media can positively contribute to the normative work required for prospective accountability, in conjunction and in dialog with other formal accountability forums (e.g. political). Journalistic practices, such as investigative reporting, amplification of responsibility assignments, solutions journalism, and editorials can help establish or spread norms that in turn support prospective accountability. Moreover, media can play a role in supporting future-oriented cognition about prospective impacts, creating opportunities to draw in the public and contribute to more interactive norm development processes.

In sum, this chapter provides a comprehensive overview of the concept of algorithmic accountability and explores the complexities and challenges in holding the human actors in algorithmic systems accountable. Given the challenges outlined, it highlights the importance of prospective accountability and develops the role of the media in shaping norms and contributing to accountability processes in various ways.

Algorithmic Accountability

To better grasp algorithmic accountability[1] I begin with some definitions for our two main concepts: algorithms and accountability. An *algorithm* can be defined as a series of steps that is undertaken in order to solve a particular problem or to accomplish a defined outcome. Engineers might design and develop algorithms as technical artifacts implemented on computers for addressing informational tasks like ranking and classification. Algorithms implemented by computers can exert considerable power as they are integrated in decision-making tasks and are scaled up to impact many people. But focusing narrowly on a technical artifact can distract from the bigger picture that algorithms exist as part of larger sociotechnical systems

[1] In the phrase "algorithmic accountability" the "-ic" suffix is used to mean "relating to", thus, "accountability relating to algorithms".

which include not only the technical artifact, but also myriad social actors, their practices, and relationships. An algorithmic system thus incorporates a range of influences from engineers, users, organizations, and other stakeholders in the social system. This sociotechnical understanding of algorithms is a crucial premise for algorithmic accountability, since, owing to the status of algorithmic technologies as moral entities but not moral agents (Johnson, 2006), it is ultimately people who must be morally responsible for the behavior of algorithmic systems.

Accountability is a concept which has been explored by a range of literature across different disciplines. A dictionary definition of accountability refers to the idea of "answerability"[2]—an actor is accountable to the extent that they feel an obligation or willingness to provide a statement explaining their conduct, as may relate to the reasons, causes, or motives for some state of affairs. Descriptions in the literature further link the concept to moral responsibility (Nissenbaum, 1996; Johnson & Mulvey, 1995; Bivins, 2006), drawing connections to causality, faulty actions, role assignments, or blame, i.e. to be "accountable for" some outcome entails some responsibility for that outcome or of the actions that led to it. We can thus meaningfully distinguish at least two meanings: (1) accountability as providing explanation, and (2) accountability as being held responsible for something (Porter et al., 2022). To deepen the second sense we can also differentiate between responsibility for an *outcome* and responsibility for implementing a *process* towards an outcome, that is "outcome accountability" versus "process accountability" (Patil et al., 2014). This distinction can be mapped to ethical perspectives, both consequentialist (i.e. responsibility for outcome) and deontological (i.e. duty to see to it that an action is taken) where "duties require specific performance of certain actions, responsibilities allow agents to choose between alternative actions having the same overall consequences" (Goodin, 1986). Another dimension of responsibility refers to the temporal perspective it takes where retrospective "backward-looking responsibility" is about taking responsibility when something bad has happened (e.g. blame), and prospective "forward-looking responsibility" is about taking responsibility to see to it that a state of affairs comes about in the future (Fahlquist, 2017), which can also imply "potential blame … where the agent who was ascribed responsibility did not achieve the expected result." (Fahlquist, 2006). Johnson argues that prospective accountability "involves practices that inform and remind individual or institutional actors that they are expected to behave in certain ways" and may ask actors to account for whether "they are adhering to rules or fulfilling their responsibilities or doing what should be done to prevent untoward events from happening in the future" (2011). Accountability can therefore correspond not only to a "breach of a duty that caused a certain negative consequence" but also to "not properly discharging a forward-looking responsibility" (i.e. a responsibility to "see to it") (van de Poel et al., 2012). As a fundamentally forward-looking activity, prospective

[2] https://www.merriam-webster.com/dictionary/accountability and https://www.merriam-webster.com/dictionary/accountable

accountability also aligns conceptually with the notion of anticipatory ethics and governance (Brey, 2017; Guston, 2014).

Writing in the discipline of public sector accountability, Bovens provides an influential definition of accountability as "a relationship between an actor and a forum, in which the actor has an obligation to explain and to justify his or her conduct, the forum can pose questions and pass judgment, and the actor may face consequences" (2007). Forums may include political (e.g. a congressional hearing), legal (e.g. a courtroom), professional (e.g. a professional society), or as I will elaborate more in this chapter, the media. Further bringing to the fore the normative dimension of accountability, Lechterman adds that accountability "involves some agent accounting *to* some other agent *for* some state of affairs *according to* some normative standard" (2020). Accountability can therefore be seen as a relational concept between agents (or between an agent and a forum/audience) where information is exchanged with respect to explaining some behavior with reference to some normative expectations for that behavior. Finally, there is a possibility for consequences based on an actor's explanation of their action (or inaction). Although actual moral responsibility should only accrue to moral agents (i.e. people and not artifacts) we might still reluctantly speak of an "accountable system" as one that is designed to provide the necessary information to enable the accountability of people (Bovens, 2010). In extending this line of definition to the domain of algorithmic systems, Wieringa emphasizes five elements: actor, forum, perspective (e.g. critiquing power vs. stimulating societally desirable outcomes), account (ex ante, ex post, and in media res), and consequences, which operate to give rise to a networked account of a system (2020).

Challenges to Algorithmic Accountability

Given the above definitional background, I will now elaborate some of the challenges to providing for algorithmic accountability in systems. In particular, I'll problematize algorithmic accountability by elaborating three overarching challenges: the sociotechnical complexity that arises from networks or actors (both human and algorithm), the information deficits that plague our ability to understand and interpret such complex systems, and the expectation setting that is crucial to provide in order to sustain accountability relationships.

Sociotechnical Complexity

The sociotechnical understanding of algorithms is important for accountability because it makes clear that technical artifacts are woven through and through with the intentions of human actors, including designers, data creators and annotators, maintainers, operators, users, and others. These actors form a complex actor network including both technical actants and human actors with connections and interactions that may give rise to emergent behavior, and where the behavior of the overall system is not easily determined by the behavior of the individual actors in the system (Holland, 2014). Interactions may be non-linear with feedback loops that amplify or dampen behaviors in non-intuitive ways and where small changes in initial conditions can result in larger impacts downstream. One aspect of sociotechnical complexity has been termed the "many hands" problem (Nissenbaum, 1996), which is that when you have many individuals (e.g. designers, developers, etc.) contributing to a sociotechnical artifact it can frustrate attempts to assign moral responsibility (or for actors to accept responsibility under unclear and uncertain conditions). This is further exacerbated when you have a network of both human and non-human actors where the non-human actors may still have causal agency, that is, the ability to initiate actions that cause downstream behavior in the system. The issue of many hands is compounded because the human hands that also contributed to setting up an autonomous system only exert control indirectly and non-continuously during design and maintenance cycles. Ibo van de Poel et al. (2012) argues that there are three main conditions necessary to hold an agent accountable: capacity, causality, and wrongdoing. Sociotechnical complexity, including the many-hands issue, appears to challenge the ability to establish causality as it clouds the lines of impact within the network of relationships, making it difficult to pinpoint who exactly is answerable for some wrongdoing.

A consequence of the artificial nature of sociotechnical systems and of the various human intentionalities embedded in such systems is that they can be unstable, indeterminate, and therefore difficult to predict. These systems are subject to human intentional agency and can change at almost any time in how they are designed, operated, or used by people. Making sense of social systems is especially challenging due to the behaviors of individual human actors, who may be swayed by external factors (including other actors or technological artifacts) but also possess the freedom to make their own decisions. These systems further exhibit instability due to the probabilistic behaviors exhibited by machine-learned models and through practices like A/B testing which deliberately introduce randomness into the experiences of users. The statistical nature of machine learning systems may be impacted both by natural stochasticity as well as human influences in the modeling process. For instance, the choice to optimize for statistical false negatives in contrast to false positives can be a deliberate one with moral implications. The non-determinism of human behavior in light of agency, combined with the probabilistic behavior of computational artifacts, a mixture of inherent and human-influenced statistical uncertainty, challenges the ability of a forum to know the conditions under which an outcome came about.

The various dimensions of sociotechnical complexity can be particularly problematic for accountability in the sense of finding those actors in the network that should face consequences. Consider the case of a harmful message on social media going viral; it may be the product of many users interacting with the message (e.g. liking, re-sharing) as well as a social media algorithm amplifying the message to even more users, a positive feedback loop. Both the users interacting with the message as well as the algorithm promoting the message could be said to share responsibility for the virality (in the sense of causally contributing to it). But it's unclear how to apportion that responsibility over the perhaps hundreds or thousands of users who have interacted with the system, as well as the engineers who designed the system to react the way it did, or learn from the data made available to it in the ways that it did. Assigning responsibility in such situations of collective or distributed action is difficult, particularly if the outcome of interest emerged from a complex chain of events between many actors, some of them technical artifacts with causal agency (Johnson & Verdicchio, 2018), but none of which have an overview or complete understanding of how their own actions might cause the complex system to behave in a particular way. Indeed, it is hard to conceive of a path forward here, aside from carefully parsing the actor-network of stakeholders, including their various relationships, and then surfacing information about their individual capacity, causality, knowledge, freedom, and wrongdoing in a specific case, tracing those relationships over time as outcomes emerge from the system. Accountability in the sense of answerability can still act as a supporting mechanism here, as it can increase the stock of knowledge about important factors that may then be used to flesh out the other preconditions that could indicate accountability in the sense of responsibility for consequences.

Information Deficits

If we focus on Boven's definition of accountability as "a relationship between an actor and a forum, in which the actor has an obligation to explain and to justify his or her conduct…" (2007) it becomes clear that there are informational prerequisites that must be met to enable the observation and understanding of the conduct of the actor. Closing information gaps is crucial to reducing uncertainty about system behavior. These information needs are largely captured by the idea of *transparency* (Diakopoulos, 2020). Transparency should be seen as a necessary but insufficient criterion for accountability; it would be difficult without at least some transparency to know enough about a sociotechnical system to request an explanation from an actor or assign responsibility for consequences. Different accountability forums may have different informational needs and there is no one-size-fits-all approach to transparency, nor should it be seen as a binary of "transparent" vs. "not transparent". Transparency information can be distinguished by whether it is demand-driven (i.e. a forum demands information), proactive (i.e. an actor decides to share information of their own volition), or forced (i.e. leaked or externally audited) (Meijer, 2014).

In light of the previous section on sociotechnical complexity, an essential component of transparency is in elaborating the various actors (both human and technical) and their relationships. Of keen importance is the degree and type of human involvement in the system, again because those intentionalities underpin the call for moral responsibility. People here may include designers, developers, maintainers, users, as well as others in the supply chain such as data annotators. Wieringa further delineates a set of information on the social side of the sociotechnical system involving the who, what, where, when, and why of an algorithmic system: who is developing and using a system, what was created exactly, where it's being developed and deployed, when was the system developed, maintained, and terminated, and why was it needed in its proposed shape (2020). Data too is a core component of algorithmic systems and information about how data was sourced, filtered, biased, and annotated can be relevant to tracing responsibility in a system. For machine-learned models transparency into the modeling process and into evaluations of model inferences can also be helpful (Diakopoulos, 2016).

As we delve into information disclosure about the technical components in the system, we must also confront issues of technical explainability and interpretability of potentially very complex systems: are there clear ways to understand an algorithm's behavior? In some cases, the behavior of a model may be difficult to explain in intuitive or direct ways, such as may be the case for some neural network models. This can threaten the ability of an accountability forum to assess wrongdoing of the technical component within the larger system, since an explanation could, for instance, offer legitimate excuses for an outcome that might shift responsibility (blame). For instance, if a model's output caused direct harm, and that output could be explained as a result of faulty input data, the blame for the outcome could then be reasonably shared with the data provider. Techniques of auditing have emerged as powerful approaches for characterizing system behavior by looking at the relationship between inputs and outputs to algorithms (Diakopoulos, 2015; Raji et al., 2020). Explainable AI approaches have also sought to develop technical approaches that are more easily interpretable, such as by using particular models where feature weights can be articulated, or logical rules can be used to provide a human-understandable explanation.

Beyond the issues of technical explainability, there are a range of pragmatic challenges to bridging information deficits for accountability forums (Diakopoulos, 2020; Busuioc, 2020). Sociotechnical systems may include informational barriers around some parts of the system, such as with governmental or organizational boundaries. Parts of the system can therefore be difficult to observe for a range of reasons, such as organizational competitiveness and reluctance to share information (e.g. to protect trade secrets around proprietary technology, or to ensure confidentiality of sensitive data), or legal statutes that specify what information may be requested (or not). Privacy considerations may place further limits on the information made available about a system. Ultimately, all of the transparency information must be disclosed in an understandable way that aligns with the needs and goals of the accountability forum; too much information may be as unproductive as too little.

Expectation Setting

For there to be accountability there must have been some negative outcome connected to a behavior or action taken by a moral agent that warrants a call for explanation, where the agent feels obligated to respond, and where the agent may be assigned blame if a sufficient response is not provided. There must have been some wrongdoing (van de Poel et al., 2012) which also means there must have been some expectation for appropriate or correct behavior or outcome. So where do the expectations for appropriate outcomes or behavior come from? How should those expectations get set and potentially updated over time? Expectations for behaviors might come from formalizations such as laws, legally enforceable policies, or charters (e.g. fundamental rights documents), or less formally through social norms, defined as the "prevailing codes of conduct that either prescribe or proscribe behaviors that members of a group can enact" (Lapinski & Rimal, 2005). Social norms provide a basis for a group to mutually understand the grounds for accountability, of "helping to support rational presumptions about not only what demands will be known to others, but also which demands will be justified to them" (van Schoelandt, 2018). Norms can also be distinguished as *descriptive* or *injunctive,* where, "descriptive norms provide information about what is done, and injunctive norms indicate what ought to be done" (Lapinski & Rimal, 2005). Injunctive norms of interest to questions of accountability might include professional norms around how algorithmic systems should be designed and developed, as well as professional norms around how technological tools ought to be used in appropriate and responsible ways.

Questions of expectation setting will likely need to be answered in context-specific ways. The norms for professional ethical behavior vary from domain to domain, suggesting that engineers may also need domain-specific understandings in order to build algorithmic systems that are in-line with those expectations (Diakopoulos et al., 2023). Different social contexts may also trigger different interpretations of wrongdoing, and therefore expectations for answerability. For instance, different social expectations around fairness may be activated when considering an algorithm used in a high-stakes domain like criminal justice in comparison to a lower-stakes context of a corporation charging different prices to different customers for a nonessential good. In public administration domains, expectations for transparency and interpretability of algorithms may be heightened in order to maintain procedural justice and even legitimacy (Diakopoulos, 2016; Busuioc, 2020). Finally, the accountability forum may also define the expectation for behavior around accountability as answerability; an actor may be compelled to respond in certain legal forums, whereas in an informal forum like the news media they may only feel compelled to respond because there is a social norm to do so.

Although it may be difficult to encode or impose a uniform set of expectations for algorithmic systems across many disparate contexts, a starting point might be adherence to fundamental rights provisions such as those expressed in the UN Declaration of Human Rights, the EU Charter of Fundamental Rights, or other such consensus human rights documents. Domain specific principles for ethical AI have also proliferated (Hagendorff, 2020), including from various professional societies, and have been critiqued due to their vagueness and difficulty in providing actionable advice (Mittelstadt, 2019). Translations of high-level abstract ideas into concrete actions are sorely needed to help guide the expected behavior of people in algorithmic systems such as designers, developers, and operators, as well as to define the expected outcomes of systems. In the EU there is activity towards formally regulating AI, which codifies expectations for providers and deployers of AI systems based on their perceived level of risk in society (Helberger & Diakopoulos, 2022). Expectations for behavior and outcomes coming from principles or formal policy, can then in turn become the basis for calls for accountability. An underlying challenge to algorithmic accountability is therefore the normative work. Some of this normative work has been happening for at least the last ten years (and perhaps longer if we consider a longer history of ethics and technology), both in the context of academic research communities like conferences for Fairness, Accountability and Transparency (FAccT) and AI Ethics and Society (AIES) as well as in the news media (Diakopoulos, 2015), as I will explore further below. In some cases, this normative work will simply feed into publicly understood social norms and expectations of behavior, but it should also inform formalizations of expected behavior through enforceable standards such as those being developed by international standards bodies.

Prospective Algorithmic Accountability: Towards a Role for News Media

The challenges outlined in the last section are all interconnected: sociotechnical complexity combined with various information deficits is a difficult bramble that serves to obfuscate chains of causality in real systems, confounding attempts to assign blame and in some cases to establish clear expectations for behavior. While retrospective accountability (i.e. with associated notions of blame and tracing cause) has often dominated discussions of algorithmic accountability, and for important reasons including the rendering of justice for individuals who experience direct harms from such systems, here I propose that prospective notions of accountability offer an important complementary path to address the challenges to accountability introduced by sociotechnical complexity and information deficits. To enable prospective accountability, I argue that news media can play an important role in addressing the challenge of expectation setting as well as by helping to close information gaps and parse sociotechnical complexity.

Prospective accountability again refers to the practice of assigning responsibilities to actors such that they are expected to behave in particular ways and take forward-looking responsibilities to see to it that a certain state of affairs is obtained (or not), and therefore that those actors may be answerable for the degree to which they fulfill those responsibilities. Prospective accountability is primarily about arranging the social side of the sociotechnical system to anticipate and then mitigate potential risks of the system, however, there are also opportunities to steer technical artifacts towards supporting more accountable systems, such as through improved explainability. Instead of focusing on causal blame for some bad outcome, a prospective approach identifies actors in the system and assigns them specific responsibilities for avoiding bad outcomes and ensuring good outcomes. An information regime can then be established so that these responsibilities can be monitored according to detailed information disclosures. However, for a regime of prospective accountability to be successful it must also address the difficult normative challenges surrounding expectation setting. A considerable amount of normative work needs to be done up-front to (1) establish the states of affairs that should be obtained (or not) in light of an algorithmic system that may be changing rapidly, and then (2) to assign forward-looking responsibilities to actors that are in a position (and have a capacity to act) to reliably see to those outcomes.

The first issue relates closely to the task of impact assessment with associated challenges of identifying real harms that may plausibly come about from the system, developing ways to reliably detect and measure those harms, and accessing various forms of expertise and impacted communities in identifying impacts (Metcalf et al., 2021). Metcalf and co-authors succinctly describe the challenge: "Organizations responsible for algorithmic systems can only be held accountable for the impacts that can be foreseen by their [impact assessment] processes and not the full set of possible outcomes that their system might produce" (2021). The second issue relates more to the idea of responsibility analysis (Fahlquist, 2017), which may involve reasoning through the sociotechnical network to identify actors that are in position (i.e. have capacity) to address the impacts in substantive ways. For instance, in the domain of synthetic media (i.e. deepfakes) that may purport to show people doing or saying things that they didn't, responsibility could be assigned to technology providers (e.g. by developing detection tools) or people who may encounter such deepfakes (e.g. by taking responsibility for educating themselves) or other stakeholders such as journalists that might be in a position to take action through factchecking or debunking (Diakopoulos & Johnson, 2021).

For prospective accountability to flourish, an important open question that needs to be answered is who should be responsible for creating the normative backdrop against which algorithmic systems can be meaningfully approached? One possibility that has been explored recently is to engage technology researchers in anticipating the potential harms of the technologies they are creating. For instance, for the NeurIPS conference in 2020, a top-tier conference in AI, authors were asked to include a broader impacts statement as part of the paper. As the conference call for papers indicated, "In order to provide a balanced perspective, authors are required to include a statement of the potential broader impact of their work, including its ethical aspects and future societal consequences. Authors should take care to discuss both positive and negative outcomes." In our research on the initiative, which involved reading and thematically analyzing a sizeable sample of the statements, we suggest that such activities could encourage reflexivity, suggest alternatives for steering future research, or even contribute to minimizing negligence and recklessness in research (Nanayakkara et al., 2021). Our follow-up work on the evolution of the requirement in 2021 showed, however, that researchers don't always perceive a sense of agency in their ability to mitigate a potential harm (Liu et al., 2022). They may deny responsibility or try to assign responsibility to practitioners, the subfield culture, or to future researchers. This work highlights an important issue for prospective accountability, namely that an actor who anticipates a state of affairs may not always be in the best position to be assigned the forward-responsibility of dealing with it, or more generally may just not be willing to take it on at all.

Still, even if researchers are somewhat effective at anticipating potential states of affairs that should be obtained (or not), technical and other experts express a range of cognitive and other biases that may narrow their visions and distort their ability to project the societal impacts of their inventions (Bonaccorsi et al., 2020; Bozeman & Boardman, 2009). Given the normative dimension here we can reasonably ask whether it is too narrow a group of people. Recent work on the topic of participatory foresight (Nikolova, 2013) advances the idea of incorporating a cognitively diverse set of participants in helping to envision the future. The idea is that many perspectives, from many cognitively diverse individuals with different backgrounds, experiences, and expertise will ultimately paint a far more nuanced and faithful picture of future algorithmic impact and safety issues. If we assume that norms should reflect some kind of shared social reality it would seem necessary to extend the groups of people tasked with the responsibility for identifying the future states of affairs that should be obtained (or not). There are of course many different methods that you could imagine supporting this, such as envisioning workshops or public forums where many diverse individuals can contribute ideas or assessments of the future. In some of our recent work, we leveraged a survey method of non-experts to capture evaluations of potential negative impacts of AI systems used in government (Barnett & Diakopoulos, 2022).

Other accountability forums may also be contenders for contributing to the necessary normative work, such as political forums (e.g. parliamentary hearings or proposals), professional forums (e.g. industry working groups or research communities), and media forums (e.g. mainstream or social). Forums can also interact in interesting and useful ways in developing norms: a parliamentary proposal like the EU's AI Act may make suggestions for responsibility assignment that are then critiqued in a professional forum such as in a research community (Hedlund, 2022; Helberger & Diakopoulos, 2023). Professional forums like industry working groups may issue statements of principles (e.g. the US-ACM's statements on Algorithmic Transparency and Accountability, or Principles for GenerativeAI) which establish normative expectations for responsible technology development. Such principles statements may then get subsequently discussed in media forums or political forums which may further amplify them or seek to incorporate them into policy proposals. In a networked view of algorithmic accountability (Wieringa, 2020) the interplay between forums is a necessary feature of how accountability is ultimately rendered.

While there are contributions to be gleaned from many different forums, here I focus on the role that the news media forum may play in establishing the normative underpinnings for prospective accountability, and in closing information gaps and parsing sociotechnical complexity. News sits alongside other media, in particular social media (Brady et al., 2021), in contributing to norm development, but as a largely professionalized activity it also offers possibilities to be deliberate in constructing its role and responsibilities as a media form and as enabling a broader matrix of forums to contribute to algorithmic accountability. Although accounts of journalistic practice have long emphasized the role of journalism in watchdogging powerful institutions and holding them accountability (Kovach & Rosenstiel, 2014), it hasn't been until more recently that research has begun to establish the theoretical underpinnings of how media contribute to accountability of actors like public institutions (Jacobs & Schillemans, 2019). Media contribute to *horizontal accountability* in the sense that they have no formal power to enforce infractions from the actors that they investigate (unlike vertical accountability as might exist in an organization), but rather function as third-party instigators. Reporters leverage their abilities in information gathering, such as freedom of information requests, interviews with insiders, leaks of information, or through external audits they may undertake, in order to help identify failures in the behavior of complex algorithmic systems (Diakopoulos, 2015, 2019), and begin an accountability process that then leads to a call for explanation from an actor. Media forums wield power by drawing public attention to issues, with the consequences being largely reputational in nature. An actor who fails to provide a satisfactory account of an outcome may appear negligent in the public eye or draw the disapproval of the public for its conduct, negatively impacting its reputation.

Jacobs and Schillemans' typology of media's roles in accountability includes four distinct functions: spark, forum, amplifier, and trigger (2019). As spark, in the course of ordinary reporting (i.e. "just asking questions") the news media may cause organizations to reconsider their behavior or role in a process. As forum, the media act as a space where investigations uncover unwanted behaviors leading to critical questions (i.e. the traditional watchdogging role). Media may also reinforce the efforts of other accountability forums, such as by drawing additional attention to congressional hearings, acting as an amplifier. As I'll elaborate more below, this role of amplifier is a crucial component of the media's impact on spreading social norms. The last role, trigger, is a sort of an extension of a spark, where the media contributes more fully to enabling other accountability forums by directly producing relevant information that spurs formal accountability. These roles are not mutually exclusive, as, for instance, the media can act as a forum that then triggers other more formal accountability forums too. In general, the information produced and publicized in media, whether as a spark, forum, amplifier, or trigger can support other accountability forums. These various roles that the media forum can take on tend to emphasize definitions of accountability as answerability or as supporting retrospective accountability (i.e. blame assignment). Here I propose that the media can also play a fifth role that is aligned with the normative work required to enable a prospective accountability regime.

As I indicated above, there are a couple areas where normative work is needed, including both establishing the outcomes that should be obtained (or not), and then assigning responsibilities to actors to see to those states of affairs. An underlying pre-condition for any of this work is also to establish norms for accountability in the sense of answerability, both as it facilitates closing information gaps, but also because it is diagnostic in terms of what actors may be ready to accept certain forward-looking responsibilities. In the course of reporting, journalists apply various news values (e.g. proximity, timeliness, controversy, surprise, reference to elites, etc.) to help identify what events in the world may warrant development into news items (Harcup & O'Neill, 2016). These news values already reflect norms amongst professional journalists about what kinds of outcomes or behaviors of actors may be detrimental and therefore warrant public scrutiny. Even in the course of their daily work reporting on technology and its mishaps, journalists therefore set the stage for establishing norms amongst the broader public for what impacts or outcomes are worthy of attention. This aligns with the known agenda-setting function of the media. By selecting impacts to report on for accountability, media begins to legitimize those impacts. The media thus plays a crucial role in establishing the norms for what outcomes should be obtained (or not) by algorithmic systems, though it's important to note that the choice of impacts that the media focuses on may not be comprehensive and is itself influenced by professional norms and therefore also subject to critique.

Upon identifying a story where a journalist thinks there has been a bad outcome (e.g. algorithmic discrimination), or bad behavior by an actor in the system (e.g. a user that misconfigured a system or neglected best practices in design or development) they can then seek accounts to help explain that bad outcome or behavior. This entails the journalist engaging in thorough stakeholder mapping of the complex system to see who *might* be answerable for the outcome, and then contacting those actors to see if they're able to explain the outcome. By enumerating stakeholders and their various relationships and interconnections, journalists articulate what the complex sociotechnical system looks like and what actors in that system could take responsibility. Through descriptions of what the actors in the system are currently doing to avert bad outcomes, journalists communicate descriptive norms. By asking certain actors for explanations (e.g. a tech developer or data annotation provider), journalists are also subtly suggesting that there could be a social expectation that the actor may need to answer for some outcome or action/inaction. Some actors may not respond to requests for explanations, though by including these gaps in their article ("i.e. XYZ did not respond to requests for comment"), journalists may also subtly signal to the public a belief that those actors maybe *should* have provided an account. Journalists can also query other stakeholders in the system (or experts who study the system such as academics, industry insiders, or representatives of professional organizations) to ask them who they think ought to be responsible for some outcome, thus contributing to the development of injunctive norms. Through these various routes journalists are thus able to interrogate a sociotechnical system and *audition* social expectations that may be candidates for norm development around forward-looking responsibilities. Media is a forum where accountability in all its forms—answerability, blame, and forward-looking—are in the making as descriptive and injunctive norms are explored via reporting on prior events and the groundwork for the broader deliberation of forward-looking responsibility assignments is laid.

In his study of the Chicago Tribune's coverage of the capital punishment reform in Illinois in the early 2000s, Ettema elaborated a two-step approach, with investigative journalism serving to establish the issue, and then editorials doing the work of amplifying reforms to the system (i.e. potential responsibility assignments) that had been suggested by a commission the government set up to examine the issue (2007). We see here the interconnection of forums as media informs formal accountability processes that then feed back into other media processes. Among the many prescriptions made by the commission the newspaper's editorials "highlighted key prescriptions … and defended them against criticisms" thus focusing attention on the responsibility assignments they deemed more warranted or efficacious (Ettema, 2007). While this is a rather subtle approach towards shaping norms, news media can also engage in somewhat more direct ways in establishing responsibility assignment through editorials or op-eds that argue explicitly for particular actors to take responsibility to see to it that certain issues are attended to. One recent example published by Wired Magazine,[3] argues for a role of government actors in reigning in AI systems in society. Another Wired example proposes a modification to red-teaming during the AI development process to ensure more resilience to institutions that may be negatively impacted.[4] Yet another example, published by Noema Magazine, calls on industry to take responsibility for exploitative labor practices in the data annotation supply chain for algorithmic systems, calls for those with funds to provide those funds to support research on labor exploitation, and calls for journalists to draw clear connections "between labor exploitation and harmful AI products".[5] All of these examples can be seen as normative auditions for responsibility assignments to see to it that certain harms of algorithmic systems are reduced moving forward, or in some cases to adapt certain processes (like red teaming) to reduce future impacts.

Solutions journalism is another strand of work in media that may productively help to build norms to support prospective accountability. Sometimes also referred to as constructive journalism, solutions journalism has been defined as "reporting on responses to social problems" (McIntyre, 2019). Solutions journalism does not call on reporters to devise their own solutions but rather to orient stories around how to make progress on addressing issues and engage sources in the course of their reporting on potential solutions. It is seen as developing a more positive vision for action rather than focusing on negativity (e.g. blame for wrongdoing) and could thus be seen to align with prospective rather than retrospective accountability. Solutions journalism explores actions and responses to harms that appear to be effective at addressing those harms. This again may be an indirect way to suggest

[3] https://www.wired.com/story/the-white-house-already-knows-how-to-make-ai-safer/

[4] https://www.wired.com/story/red-teaming-gpt-4-was-valuable-violet-teaming-will-make-it-better/

[5] https://www.noemamag.com/the-exploited-labor-behind-artificial-intelligence/

norms for responsibility as in "here is a case where some action A appears to be addressing harm X". Solutions framed stories have been empirically shown to encourage audiences to report more motivation to take positive actions in response to an issue (Baden et al., 2018).

Another way that media contributes to the normative work to support prospective accountability relates to its broader role towards supporting future-oriented cognition in society (Diakopoulos, 2022). Future-oriented cognition describes how people engage in setting goals, planning ways to reach those goals, and mentally simulating situations that may occur. The media engages in such processes through activities such as predictive journalism (e.g. election predictions, COVID vaccination predictions, etc.), which leverage algorithms to extrapolate information into the future, develop and communicate scenarios about what might happen, or explore counterfactual conditions influencing events. This type of work can also make salient how a simulated algorithmic system may create certain kinds of impacts. For instance, in a recent algorithmic accountability report that examined a risk scoring algorithm used in welfare benefits investigations in the city of Rotterdam, the journalists developed archetypes that the algorithm might score in particular ways.[6] As they write: "Here's how the city's welfare algorithm reduces a struggling mother's complex life into simplified data. Imagine "Sara," a hypothetical single mother of two children. She recently broke up with her partner. When one of her sons fell ill 4 months ago, she quit her job to take care of him. She now runs out of money to pay her bills before the end of each month. At their last meeting, Sara's caseworker noted in her file that "clothing, makeup, and/or hairstyle are unsuitable for job application" and that she was struggling to deal with setbacks because of a lack of flexibility. Even these subjective observations influence the fraud algorithm." This hypothetical extrapolation makes clear that there are potential outcomes of using the algorithm that create harm for an individual, and also insinuates that there may be improper features used by the algorithm. Thus, by projecting potential harms and the conditions in which those harms might arise, this type of journalism both identifies future states to avoid, and also subtly suggests a responsibility for developers (i.e. related to feature choice) that would avert bad outcomes.

By publishing prospections of potential outcomes from complex systems, including algorithmic predictions, media can also encourage engagement by the broader public in participating in the normative work needed to support responsibility assignment. In our study of comments in response to COVID-19 predictive journalism, we found that commenters often referred to actions that they thought would address projected outcomes: "commenters were engaged in a group planning activity where particular actions and responsibilities are articulated for future action" (Allaham & Diakopoulos, 2022). In other words, the media, by framing outcomes and creating a space for discourse can also support more inclusive social processes necessary to support norm development and responsibility assignment. In particular, by allowing non-elites to contribute to the articulation of descriptive or

[6] https://www.wired.com/story/welfare-state-algorithms/

injunctive norms, or simply to express evaluations of behavior or opinions on responsibility, professional media acts as a facilitator. To close the loop, news organizations might further shepherd the process by, for instance, summarizing community responses to provide an overview, orienting discussions towards further critique, or amplify some responses that appear to be promising or have wider community support.

Of course, news media, or even media more broadly, is not a panacea for accountability. While I would argue it is an important element for enabling (prospective) accountability not only in helping to establish and socialize norms but also in parsing sociotechnical complexity and closing information gaps, it also has its weaknesses. One issue is that the professional logics of what gets selected in news media, as well as the downstream logics of algorithmic platforms which render attention according to complex feedback loops of user engagement, mean that media attention is never systematic and may not be equitably distributed. Only in a sufficiently plural media system could we expect that the biases inherent to the media system might be averaged out. In its ideal state, diverse individuals and organizations participate to constitute the broader media. Another issue is that in order to maintain its legitimacy news media must retain its independence from the entities that may be at the heart of accountability issues with AI. In particular, if media become captured by large technology corporations (such as through technological reliance), they may be less apt to explore responsibilities for violation of expected behavior of those technology corporations. The underlying concern is about impartiality: news media must try to manage their own relational biases if they are to legitimately contribute to shaping normative expectations.

Conclusions

While there has been a predominant focus in understanding algorithmic accountability through a retrospective lens, this chapter suggests that prospective accountability may be an important complement for addressing underlying challenges such as the sociotechnical complexity of algorithmic systems as well as the many information gaps that thwart blame assignment. Yet norms and expectations must also be developed to support accountability, including for prospective accountability. This includes norms for answerability, norms for establishing what is harmful and impactful that society may want to avoid, and norms for assigning responsibility to particular actors for mitigating those potential states of affair. In this chapter I have elaborated a role for the news media to play in enabling and supporting a prospective accountability regime. By helping to establish and sustain answerability, identify potential negative outcomes to avoid, and suggest and amplify responsibilities for actors that might be capable of reducing future harms, it does fundamental work to support algorithmic accountability. If certain activities were further emphasized, such as editorial writing, solutions journalism, predictive journalism, and the facilitation of interactive public engagements it could be even more impactful in guiding the way for prospective accountability.

References

Allaham, M., & Diakopoulos, N. (2022). *Predicting COVID: Understanding audience responses to predictive journalism via online comments*. New Media & Society.

Baden, D., McIntyre, K., & Homberg, F. (2018). The impact of constructive news on affective and behavioural responses. *Journalism Studies, 20*(13), 1940–1959. https://doi.org/10.1080/1461670x.2018.1545599

Barnett, J., & Diakopoulos, N. (2022). Crowdsourcing impacts: Exploring the utility of crowds for anticipating societal impacts of algorithmic decision making. Proc. AI Ethics and Society (AIES) https://doi.org/10.1145/3514094.3534145

Bivins, T. H. (2006). "Responsibility and Accountability" in Ethics in Public Relations: Responsible Advocacy, 19–38. https://doi.org/10.4135/9781452204208.n2

Bonaccorsi, A., Apreda, R., & Fantoni, G. (2020). Expert biases in technology foresight. Why they are a problem and how to mitigate them. *Technological Forecasting and Social Change, 151*, 119855. https://doi.org/10.1016/j.techfore.2019.119855

Bovens, M. (2007). Analysing and assessing accountability: A conceptual framework. *European Law Journal, 13*(4), 447–468. https://doi.org/10.1111/j.1468-0386.2007.00378.x

Bovens, M. (2010). Two concepts of accountability: Accountability as a virtue and as a mechanism. *West European Politics, 33*(5), 946–967. https://doi.org/10.1080/01402382.2010.486119

Bozeman, B., & Boardman, C. (2009). Broad impacts and narrow perspectives: Passing the Buck on science and social impacts. *Social Epistemology, 23*(3–4), 183–198. https://doi.org/10.1080/02691720903364019

Brady, W. J., McLoughlin, K., Doan, T. N., & Crockett, M. J. (2021). How social learning amplifies moral outrage expression in online social networks. *Science Advances, 7*(33), eabe5641. https://doi.org/10.1126/sciadv.abe5641

Brey, P. (2017). Ethics of emerging technology. In S. O. Hansson (Ed.), *Methods for the ethics of technology: Methods and approaches* (pp. 175–192). Rowman & Littlefield International.

Busuioc, M. (2020). Accountable artificial intelligence: Holding algorithms to account. *Public Administration Review, 81*, 825–836. https://doi.org/10.1111/puar.13293

Diakopoulos, N. (2015). Algorithmic accountability: Journalistic investigation of computational power structures. *Digital Journalism, 3*(3), 398–415.

Diakopoulos, N. (2016). Accountability in algorithmic decision making. *Communications of the ACM, 59*(2), 56–62.

Diakopoulos, N. (2019). *Automating the news: How algorithms are rewriting the media*. Harvard University Press.

Diakopoulos, N. (2020). Transparency. In M. Dubber, F. Pasquale, & S. Das (Eds.), *Oxford handbook of ethics and AI*.

Diakopoulos, N. (2022). *Predictive journalism: On the role of computational prospection in news media*. Tow Center for Digital Journalism.

Diakopoulos, N., & Johnson, D. (2021). Anticipating and addressing the ethical implications of deepfakes in the context of elections. *New Media & Society, 23*(7), 2072–2098. https://doi.org/10.1177/1461444820925811

Diakopoulos, N., Trattner, C., Jannach, D., Costera Meijer, I., & Motta, E. (2023). Leveraging professional ethics for responsible AI. *Communications of the ACM* (CACM).

Ettema, J. S. (2007). Journalism as reason-giving: Deliberative democracy, institutional accountability, and the news media's mission. *Political Communication, 24*(2), 143–160. https://doi.org/10.1080/10584600701312860

Fahlquist, J. N. (2006). Responsibility ascriptions and vision zero. *Accident Analysis & Prevention, 38*(6), 1113–1118. https://doi.org/10.1016/j.aap.2006.04.020

Fahlquist, J. N. (2017). Responsibility analysis. In S. O. Hansson (Ed.), *Methods for the ethics of technology: Methods and approaches*. Rowman and Littlefield International.

Goodin, R. E. (1986). Responsibilities. *The Philosophical Quarterly, 36*(142), 50. https://doi.org/10.2307/2219309

Guston, D. H. (2014). Understanding “anticipatory governance”. *Social Studies of Science, 44*(2), 218–242.

Hagendorff, T. (2020). The ethics of AI ethics: An evaluation of guidelines. *Minds and Machines, 30*(1), 99–120. https://doi.org/10.1007/s11023-020-09517-8

Harcup, T., & O’Neill, D. (2016). What is news? News values revisited (again). *Journalism Studies, 23*(1), 1–19. https://doi.org/10.1080/1461670x.2016.1150193

Hedlund, M. (2022). Distribution of forward-looking responsibility in the EU process on AI regulation. *Frontiers in Human Dynamics, 4*, 703510. https://doi.org/10.3389/fhumd.2022.703510

Helberger, N., & Diakopoulos, N. (2022). The European AI act and how it matters for research into AI in media and journalism. *Digital Journalism*.

Helberger, N., & Diakopoulos, N. (2023). ChatGPT and the AI Act. *Internet Policy Review, 12*(1).

Holland, J. H. (2014). *Complexity: A very short introduction*. Oxford University Press.

Jacobs, S. & Schillemans, T. Media and public accountability: typology and research agenda. in Media and Governance. (Polity Press, 2019). https://www.cambridge.org/core/books/abs/media-and-governance/media-andpublic-accountability-typology-and-research-agenda/79F79D5D5F9A3758B7234787378A83F4

Johnson, D. G. (2006). Computer systems: Moral entities but not moral agents. *Ethics and Information Technology, 8*(4), 195–204. https://doi.org/10.1007/s10676-006-9111-5

Johnson, D. G. (2011). Software agents, anticipatory ethics, and accountability. In G. E. Marchant et al. (Eds.), *The growing gap between emerging technologies and legal-ethical oversight*. The International Library of Ethics, Law and Technology, 7.

Johnson, D. G., & Mulvey, J. M. (1995). Accountability and computer decision systems. *Communications of the ACM, 38*(12), 58–64. https://doi.org/10.1145/219663.219682

Johnson, D. G., & Verdicchio, M. (2018). AI, agency and responsibility: The VW fraud case and beyond. *AI & SOCIETY, 34*(3), 1–9. https://doi.org/10.1007/s00146-017-0781-9

Kovach, B., & Rosenstiel, T. (2014). *The elements of journalism* (3rd ed.). Three Rivers Press.

Lapinski, M. K., & Rimal, R. N. (2005). An explication of social norms. *Communication Theory, 15*(2), 127–147. https://doi.org/10.1111/j.1468-2885.2005.tb00329.x

Lechterman, T. M. (2020). The concept of accountability in ai ethics and governance. In J. Bullock et al. (Eds.), *The Oxford handbook of AI governance*. Oxford University Press.

Liu, D., Nanayakkara, P., Sakha, S., Abuhamad, G., Blodgett, S. L., Diakopoulos, N., Hullman, J., & Eliassi-Rad, T. (2022). Examining responsibility and deliberation in AI impact statements and ethics reviews. In *Proceedings of the Conference on AI, Ethics, and Society* (AIES).

McIntyre, K. (2019). Solutions journalism: The effects of including solution information in news stories about social problems. *Journalism Practice, 13*(8), 1029–1033. https://doi.org/10.1080/17512786.2019.1640632

Meijer, A. (2014). Transparency. In M. Bovens, R. E. Goodin, & T. Schillemans (Eds.), *The Oxford handbook of public accountability*. Oxford University Press.

Metcalf, J., Moss, E., Watkins, E. A., Singh, R., & Elish, M. C. (2021). Algorithmic impact assessments and accountability. In *Proceedings of the 2021 ACM conference on fairness, accountability, and transparency* (pp. 735–746). https://doi.org/10.1145/3442188.3445935

Mittelstadt, B. (2019). Principles alone cannot guarantee ethical AI. Nature Machine Intelligence, 1, 501–507.

Nanayakkara, P., Hullman, J., & Diakopoulos, N. (2021). Unpacking the expressed consequences of AI research in broader impact statements. *Proceedings of the conference on AI, ethics, and society* (AIES).

Nikolova, B. (2013). The rise and promise of participatory foresight. *European Journal of Futures Research, 2*, 33.
Nissenbaum, H. (1996). Accountability in a computerized society. *Science and Engineering Ethics, 2*(1), 25–42. https://doi.org/10.1007/bf02639315
Patil, S. V., Vieider, F., & Tetlock, P. E. (2014). The Oxford handbook of public accountability. In M. Bovens, R. E. Goodin, & T. Schillemans (Eds.), *The Oxford handbook of public accountability*. Oxford University Press. https://doi.org/10.1093/oxfordhb/9780199641253.013.0002
Porter, Z., Zimmermann, A., Morgan, P., McDermid, J., Lawton, T., & Habli, I. (2022). Distinguishing two features of accountability for AI technologies. *Nature Machine Intelligence, 4*(9), 734–736. https://doi.org/10.1038/s42256-022-00533-0
Raji, I. D., Smart, A., White, R. N., Mitchell, M., Gebru, T., Hutchinson, B., Smith-Loud, J., Theron, D., & Barnes, P. (2020). *Closing the AI accountability gap: Defining an end-to-end framework for internal algorithmic auditing*. Proceedings of the 2020 Conference on Fairness, Accountability, and Transparency (pp. 33–44). https://doi.org/10.1145/3351095.3372873
van de Poel, I., Fahlquist, J. N., Doorn, N., Zwart, S., & Royakkers, L. (2012). The problem of many hands: Climate change as an example. *Science and Engineering Ethics, 18*(1), 49–67. https://doi.org/10.1007/s11948-011-9276-0
van Schoelandt, C. (2018). Moral accountability and social norms. *Social Philosophy and Policy, 35*(1), 217–236.
Wieringa, M. (2020). What to account for when accounting for algorithms: A systematic literature review on algorithmic accountability. *FAT* '20: Proceedings of the 2020 Conference on Fairness, Accountability, and Transparency*, 1–18. https://doi.org/10.1145/3351095.3372833.

Chapter 9
Understanding Sociotechnical Systems

Thomas M. Powers

Abstract Sociotechnical explanations can contribute to individual failure investigations and even normative critiques of an entire system, in the event that clear descriptions of causes and conditions of failure can be specified. But in many failed systems, what has happened is complicated. At present, sociotechnical systems approaches in organizational studies, engineering, and philosophy do not meet standards of clarity that are required of explanation. Further, sociotechnical system approaches so far have not been used to predict instances of failure, though it would be highly desirable if we could develop such predictions. Complications with explanation and the dim prospects for prediction lead us in the direction of new computational technologies, but these are not without risks. Progress in understanding sociotechnical systems means, in effect, that explanations and predictions would become useful for "unenhanced" human understanding, not just for computational models of sociotechnical behavior. One byproduct of this better understanding could be guidance on when to hold certain parts of AI sociotechnical systems accountable—namely, when the algorithms by which they operate could be held accountable for damages that they cause.

Keywords Sociotechnical systems · Technological failures · Accountability · Computational turn

Introduction

In normative studies of technology, the invocation of a sociotechnical system is intended to answer questions of a critical sort, for instance, whether we should embrace a technology, or work to improve it, or—failing improvement—whether

T. M. Powers (✉)
University of Delaware, Newark, DE, USA
e-mail: tpowers@udel.edu

M. Noorman, M. Verdicchio (eds.), *Computer Ethics Across Disciplines*, Philosophy of Engineering and Technology 44,
https://doi.org/10.1007/978-3-031-87974-6_9

the technology ought not to be allowed to operate in society. This invocation appeals to the social, political, and legal contexts in which the technology exists, and thus the concept of a sociotechnical system is by its nature a complicated thing. Since some of these questions call for explanations of deficient behaviors of a technology, in that they point to technological failure, a main idea behind the invocation is that we are engaged in a search for explanations because we may want to change the system, and invoking the apparatus of the sociotechnical system will help us to know how to do this. Even further, explaining technological behaviors as part of a system may have further epistemic utility; it may contribute to a richer understanding of the system as a whole, apart from its failures. But such "sociotechnical system explanations" are not themselves straightforward, as they involve several complicating factors. The parts of this "complicated thing" are heterogeneous, and relations between these parts can be opaque. Some of these relations are causal and some mere correlations; it is difficult to tease out which are which. Further, sociotechnical systems are dynamic. Especially because patterns of human behaviors, laws, and organizations are subject to change, it is difficult to see why the event to be explained emerged from the system in which it was situated and not from some earlier iteration of the system. Similarly, it is difficult to know whether the event will recur when changes to the system are introduced. Even assuming that researchers can overcome these complications, the fact that explanations are inherently retrospective limits the usefulness of the sociotechnical systems approach for understanding large, dynamic systems. Unlike physical systems, when we examine sociotechnical systems, we cannot simply assume that the future will resemble the past. Thus, a challenge emerges. On the one hand, it would be desirable to have better explanations of technologies (and their failures) in their sociotechnical contexts, and also to develop timely predictions of future behaviors in prospective implementations, so as to achieve greater safety and efficiency. On the other hand, focusing on a particular sociotechnical system is not likely to yield better explanations and predictions of different, future systems. So, progress in meeting both the explanation and prediction goals will require that we understand sociotechnical systems in a more thoroughgoing way, and thus answers to our normative questions will become clearer as this understanding improves. The prospects for developing such an improved understanding of sociotechnical systems are the focus of the present chapter.

To explain technological failures and to anticipate them as well, turning to sociotechnical systems is surely looking in the right place. But the payoff of such investigations has not fully materialized in the two main fields in which they originated in the twentieth century: organizational studies and engineering. In the section that follows we will see why this was the case. After diagnosing shortcomings with these early attempts, we will turn to the domain of current and future sociotechnical systems, where software and even artificially intelligent agents play significant roles. Our main concern with these sociotechnical systems will be the role of the "computer layer" in their behaviors, and this issue will occupy the latter part of the chapter.

One particularly illuminating kind of contemporary technological failure has to do with "algorithmic bias"—for instance, when an AI algorithm produces a recidivism prediction score for bail, parole, or sentencing considerations by a judge (Dressel & Farid, 2018), but the result is statistically biased in that it overestimates recidivism risk for a particular demographic group. How and why an AI tool does this can vary over models; sometimes the problem is in the selection of the training data, sometimes in the algorithm that was used, and sometimes in both. Whichever is the case, such an outcome certainly calls for an explanation, and supposing that one is forthcoming, we can see a further challenge for the sociotechnical systems approach to explanation. Some explanations also justify a result—this could be the case with certain AI recidivism risk tools—while other explanations are perfectly fine as explanations, but do not justify their results. These "explained but not justified" results are technological failures, and their existence suggests that additional action ought to be taken to hold someone (or something) accountable. In the final section we will consider the problem of "algorithmic accountability" as presented in Johnson (2021) as one which requires a sociotechnical explanation. Johnson's critique of algorithmic accountability sets a high bar for application of certain normative and legal concepts to technologies. But as we shall see, some kinds of AI technologies—ones that are amenable to thoroughgoing sociotechnical explanations—may clear some version of that bar in the future. In addition, the distinction between justifying and non-justifying explanations can serve as a guide as to when to pursue accountability for misbehaving algorithms and when to look elsewhere in a sociotechnical system for accountability.

Before we begin, a note about terminology is in order. In the account that follows we will understand 'engineering failures' widely, across a range of cases. Failure will be seen in cases where documented specifications (requirements of function, design, and service) are not satisfied, and other cases where technologies fall short of our social expectations—they disappoint or dissatisfy, without violating specifications. For clarity we will designate the first kind as "specification failures" and the second as "expectation failures."

Origins in Organizational Studies and Engineering

The path to an improved understanding of sociotechnical systems leads first to the origins and motivations of early attempts at sociotechnical explanation by Trist and Bamforth (1951). Looking at engineering models of failure by Leveson (2005) and philosophical accounts of sociotechnical systems (Vermaas et al., 2011) make apparent some deficiencies of explanation in all three fields. Moving beyond explanation, prediction of technological behavior enters the discussion as one response to the call for anticipatory ethics by Brey (2012a), but here too we will find that the sociotechnical systems approach will face difficulties.

Attempts to explain a failed technology force a choice of where to focus. Beginning with the sociotechnical system as opposed to the artifact or the technology proper leads one to take a broader perspective than is perhaps obvious at first glance, to see how and why a system came about and how it operates as situated. Thus, we look at people, organizations, and social context in addition to the artifact or technology design itself. In the first widely acknowledged example of such an investigation, two researchers from the Tavistock Institute in London, Eric Trist and Ken Bamforth (1951), were tasked with studying longwall coal mining in post-WWII England. The longwall method was introduced shortly after the war and nationalization of the coal mines. It was a mass-production engineering method with mechanized "coal cutters" and conveyor belts, worked by groups of 40–50 men, that replaced earlier "hand-got" coal mining methods employing small groups, usually with two members using hand tools. From an engineering perspective, the longwall method ought to have produced much more coal. In the established mines of England, it did not fail in most cases to increase production, but it was not as productive as had been projected.

Through interviews and discussions with 20 workers, coal industry psychiatrists, and management over 2 years, Trist and Bamforth concluded that social and psychological factors were responsible for the poor performance of longwall coal mining. They cited the loss of both "responsible autonomy" and adaptability of the former small workgroups when the new method was introduced. The workers also expressed a distrust of nationalization of the coal mines, though it was not clear how exactly this change affected production. Pay structures changed as a result of the longwall method, and with them the incentive to contribute to gains for co-workers was weakened. From a workflow perspective, the longwall method was vulnerable to "bad conditions" and "bad work" which created cycle stoppages, some of which workers blamed on their co-workers. Lost wages from stoppages caused further deterioration of morale, and these tensions bled over into tense social interactions in nearby towns. Staying home from work became a coping mechanism, and absenteeism increased for the miners in part because the larger size of work groups meant that the slightly diminished cohort could still function, unlike the small groups of early hand-got methods.

Attention to the "psycho-social whole" in industrial production is what set the Tavistock researchers apart, as an application of what they called "action research" (Mumford, 2006). The Institute would grow to export methods and researchers to much of the industrialized world in the coming decades, though in some fields the enthusiasm for the methods was short-lived (Davis et al., 2014). Nonetheless, the longwall method study became a centerpiece of Tavistock research. These innovations notwithstanding, it is difficult to isolate a clear instance of causal explanation in their narrative. On the whole, it sounds convincing, but limitations in the method meant that they could not show that the poor production was not due to other, undiscovered factors such as the depletion of formerly plentiful and safe coal seams during the wars. Likewise, due to cost and feasibility, they could not run a randomized controlled trial on other mines, with other coal-getting methods and workers. Nor did they establish that the detrimental effect of production with increasing work group size was linear; perhaps at groups of 100 or more overall mine production would increase. Finally, they did not consider that the poor production was not a temporary phenomenon—one perhaps easily addressed by more holidays or better pay—because as Trist and Bamforth (1951) acknowledged, the study lasted only 2 years.

In these respects, the analysis of the longwall coal-mining method by Trist and Bamforth resembles Eisenstein's monumental study of the Gutenberg movable type printing press: *The Printing Press as an Agent of Change* (1980). On her account, the fifteenth century printing press brought with it a sudden increase in availability of books, which had a profound change on the preservation and conveyance of knowledge. Important intellectual developments in the West, such as the Renaissance, the Protestant Reformation, and the scientific revolutions of the Enlightenment were all brought on by the agency of the printing press. No doubt, the declared agency of the printing press does not enter into a failure explanation—quite to the contrary. But in its claims of causal agency this account shares with the Trist and Bamforth study a narrative as opposed to logical force. They both represent methodologies that deliver on the plausibility of their respective conclusions, but do not demonstrate a causal chain which leads to them. This difference is important for explanations of some urgency or with higher stakes, where technologies impact safety. In the case of the historical significance of the printing press, there is no action that Eisenstein's readers are compelled to take, beyond having a greater appreciation of literacy and the availability of books. The coal mine managers and National Coal Board were in a somewhat more urgent situation. On the advice of Trist and Bamforth, they could try to increase the responsible autonomy of work groups and/or decrease their size. They could develop yet another mechanized production method. Any of these changes would bring costs and perhaps even new dangers to an already dangerous profession.

High-stakes explanations are typical in engineering, especially after cases of failures that lead to demands for an accident investigation by an engineering safety researcher. The work of Nancy Leveson serves as a good example of the sociotechnical systems approach to engineering safety in order to understand accident causation. Leveson (2005) finds shortcomings in traditional "event-chain" accident models because they only work well for physical device failure in simple systems when a single direct cause can be identified. This cause is typically the event most proximate to the failure event, and is "politically" acceptable, i.e., the "cure" is known and no other significant changes need be made to correct for the failure. This model does not do well at explaining accidents caused by software-intensive systems nor for failures of social aspects of the system that are contributed by organizational culture, personnel roles, regulatory lapses, poor communication, and the like. In other words, identifying technical (device) failures is politically acceptable because it does not lay blame with people, organizations, and governments. But this does not begin to tell the whole story.

Leveson uses engineering systems theory to develop a new accident model, Systems-Theoretic Accident Model and Processes (STAMP), that explains system accidents which arise from the interactions among components rather than individual component failure. "When using a systems-theoretic accident model, accidents are viewed as the result of flawed processes involving interactions among system components, including people, societal and organizational structures, engineering activities, and the physical system" (2005, p. 69). Within STAMP there is a novel account of failure causation. While in simple (event-chain) accidents we can assume a single counterfactual ("but for X, Y would not have occurred") explanation, in sociotechnical systems we must also look for causes that are contributory conditions, as well as for indirect and non-linear causation. While it would be tempting to assign this "messy" causality entirely to the social side of the system, that is not what STAMP does. In an in-depth analysis of the 1999 Titan IV/Centaur launch, which placed a satellite into a useless low orbit, Leveson finds that software control failures interacted with social failures of coordination, testing, communication, and management, and an individual worker failure to calculate a numerical value—which should have been checked and re-checked by other workers and automated systems. In sociotechnical systems, Leveson shows, failure can be messy all the way down.

Some cases of failure occur despite technology manufacturers' attempts at resolving them. Taking a cue from the STAMP model, we can identify three types of sociotechnical failure that arise from interactions between social and technical components, where technology modifications merely alter the locus of the problem. Since the mobile phone is a near-ubiquitous technology, we will describe these cases with mobile phones—though examples might be readily found in other sociotechnical systems.

1. Obfuscation Failures

Since the advent of mobile phones, there have been privacy concerns due to the amount and kind of data collected on the phone and later transmitted to and stored by service providers. Because billions of users worldwide now keep their phones with them at all times, and indeed have developed a "personalized" attachment to the phone (Srivastava, 2005), the nature of the data is quite sensitive, ranging from location, contacts, purchases, and browsing, to other app-generated data. Regarding app data, users concerned about privacy increasingly take advantage of opt-in and opt-out settings to limit privacy risk—thus giving the appearance of a technological fix to an ethical (and sometimes legal) problem. However, some of these settings solutions are far less protective than they appear. To reduce unwanted cross app tracking, in iOS 14 Apple introduced *App Tracking Transparency* (ATT), a code-enforced requirement for non-Apple apps to get users permission to track their phone activity across apps. In part this change was brought on by GDPR provisions, though Apple has also tried to establish itself as a privacy leader among the big tech companies. According to Kollnig et al. (2022), ATT has been generally effective in reducing unwanted tracking. But in response, large tracking companies (including Apple itself) developed techniques called "fingerprinting" and "cohort tracking" that allow them to link user data with data they legitimately have "behind the scenes." The outcome is that overall less data are tracked, but since linking capabilities make the data that are tracked more useful for some large companies, and users mistakenly believe that ATT diminishes the need for data tracking vigilance when using apps, user privacy may be less protected in the end. Kollnig, et al. find that a "direct result of the ATT could therefore be that existing power imbalances in the digital tracking ecosystem get reinforced" (2022, p. 515). In short, some small app companies lose out because of ATT, and users may believe falsely that they have protected the privacy of their data overall. This is a form of technological obfuscation—the appearance that a technological failure has been fixed, whereas it has just been transformed.

2. Augmentation Failures
Some technologies augment social problems that exist prior to their deployments, and thus attempts to address a specific technological contribution face difficulties in how to disentangle the respective social problem from the technology that, in part, drives it. With the introduction of mobile phones, the initial features were few; after the defining feature of mobile calling, eventually e-mail, text messaging, web browsing, geolocation, and video streaming were rolled out (Srivastava, 2005). Some less-advertised benefits of these features include the increase in citizen journalism through the producing and sharing of videos of police brutality, as in the George Floyd murder (Richardson, 2020). New uses of mobile phones also tragically seem to have encouraged cases of planned mass violence, as with livestreaming by mass shooters. One socially problematic use that has been met with different legal and technological responses is video voyeurism in the form of "upskirt" image capture (McCann et al., 2018). Early on, industry self-regulation of device settings in Japan seemed promising as an example of an industry taking responsibility for a problem, whether or not their technology had produced it. Device settings on all Japanese phones produced a shutter sound when the camera was used (Law Library of Congress, 2008). This almost-immediate change to devices was achieved by an industry association agreement and covered all phones sold in Japan. Many prefects in Japan would also end up using Anti-Nuisance Laws to prosecute cases of upskirting, but there is no publicly-available documentation of industry self-regulation nor government negotiation with industry. There is no nationwide law either. South Korea had a similar device restriction, but with documentation of the public law. In the U.K. a law was passed in 2019 that specifically addressed upskirting, but 1150 violations of that law have been recorded in the U.K. since it was enacted, and 40% of those have been children (Porter, 2023). In the U.S., in 2004 the federal Video Voyeurism Protection Act became law, but it only covers acts committed on federal lands (National Parks, etc.). While the law was intended to serve as a model for future state laws, few states have followed suit and so legal protections are predictably inconsistent (McCann et al., 2018). Indeed, some early state legal cases in the U.S. were decided in favor of the defendants on the grounds that the persons who were photographed did not have a reasonable expectation of privacy in the public spaces where the incidents occurred. Legal norms in the U.S. pull in two directions when it comes to upskirting: protecting those being photographed vs. protecting those doing the photographing. The lessons here for mobile phone image capture are that certain technological failures have been treated inconsistently, and coordination between law and industry is made difficult by national differences in laws and industry practices. The mobile phone

remains a technology that can facilitate invasions of privacy, and its features have continued to merge with antecedent behaviors and subsequent industry and legal reactions to leave a patchwork of social, business, and legal practices, depending on where one lives.

3. Overlap Failures
Technologies such as the mobile phone are often used in combination with other technologies, and in particular, with automobiles. Thus, two or more sociotechnical systems can overlap and produce results that would be difficult to explain by viewing one system in isolation. Mobile phone use is now common in all countries with widespread cellular networks and road systems, and the adverse health impacts from distracted driving are clear and alarming (Ortega et al., 2021). Long-term data show that distraction per se—considered as anything that interferes with the driver's attention—presents a significant safety risk no matter the source; any distraction competes with the cognitive workload of the driver's assimilation of information about road and traffic conditions. While mobile phone distracted driving can be lessened by hands-free applications and by laws against texting-while-driving, the huge increase in the proportion of drivers with phones suggests that this is an uphill battle. Worse still, the intensity of distraction tracks with the increase in numbers. Mobile phone distracted driving is exacerbated by a growing psycho-social phenomenon known as Maladaptive Mobile Phone Use (MMPU), or "smartphone addiction" (Rahmillah et al., 2023). Not surprisingly, mobile phone users who text, e-mail, and use applications obsessively are more likely to do so while driving, and those who display high levels of MMPU are more likely to cause safety-critical accidents. And not all of the problems created by the intersection of mobile phones and automobiles are unintentional. Remote hacking of various automobile systems has been an area of intense industry interest (and amateur hacking) for the last decade (Gitlin, 2023). Well-known security hackers Miller and Valasek (2015) accessed a 2014 Jeep and showed that physical systems such as braking and steering could be disabled remotely while the vehicle was in motion. With increasing cross-platform software shared by phones and automobiles, as well as an increase in remote computer system access from Bluetooth and NFC designed into the latter, researchers anticipate remote car hacking will become a real worry in the very near future. The problems that come with each sociotechnical system—mobile phones and automobiles—cannot be straightforwardly explained by considering each system alone, and attempts to resolve problems are unlikely to bear fruit until we have a theory of system-to-system interaction.

Philosophy and STS

While modern technologies emerged in the eighteenth century, arising out of the Industrial Revolution, what early philosophical studies of technology there were did not betray an interest in systems, for reasons that may in retrospect seem obvious.

First, early technologies were lesser in scope and complexity (Johnson & Powers, 2005), so in all likelihood the urgency to study sociotechnical systems was not yet ripe. Also, around the same time there was a reaction against system building in philosophy—captured in Nietzsche's admonition to avoid systematizers, as there is something dishonest in the pursuit—in the wake of the grandiose and overly conceptual nature of philosophical systems of Kant and Hegel. Further, there was not yet an academic sociology, as the nascent branching off of sociology from philosophy had just begun in Germany, and sociology too was one of several disciplines that was subject to the *Methodenstreit*, or "fight over method." Neo-Kantians, under the influence of Wilhelm Dilthey, saw sociology as one of the *Geisteswissenschaften*, and the prescribed methods for these disciplines were not what we would now call empirical. But technology was nonetheless subject matter for sociology, and by the turn of the twentieth century, Max Scheler and Georg Simmel had begun to investigate the role of technologies in society (Gunderson, 2017). By the middle of the twentieth century, critiques of technology combined with those of economic life, and the popular success of Marcuse's description of the "machine process" (Marcuse, 1964) produced a broader worry about modern technology as a system of social control. For Marcuse, modern industrial societies were organized around production and consumption in a way that produced false needs and new forms of control—resulting in a "smooth, comfortable unfreedom." In claiming that "political power asserts itself through its power over the machine process and over the technical organization of the apparatus" (1964, p. 3), he is directly indicting both advanced industrial capitalism and (at the time) the competing socialist planned economies. He is however surprisingly optimistic about humanity re-gaining control of the "machine process," claiming that mechanized work could become "the *potential* basis of a new freedom for man." Later philosophers of technology would be greatly influenced by Marcuse.

Concurrently, there arose in academic philosophy of science in the middle of the twentieth century a keen interest in explanation. The leading model was developed by Hempel and Oppenheim (1948) and Hempel (1965) and described explanations of the events of interest to the physical sciences as deductions from the causal laws of those sciences. This model proved not to be suitable for explanation in the social sciences, where many of the "laws" were statistical and not transparently causal. In reaction, Salmon (1971) developed a model of explanation as statistical relevance. For Salmon, explanation in the social sciences was a process of supplying information relevant to the event or phenomena to be explained, not a matter of giving an argument from which, with knowledge of causal laws, we could deduce what had happened. Note too that on the Hempel model, explanation and prediction were essentially collapsed; any explanation of an event that contained the relevant law of nature would serve as a prediction of a similar event in the future, assuming that the laws of nature did not change. Would *technological* behaviors follow the causal or the statistical model of explanation? Twentieth-century philosophers of science would never broach the question. As fertile as this discussion could have been for a philosophy of technology, there were at the same time renewed countervailing trends against systemizing. A major influence in the physical sciences (as seen from the perspective of philosophers of science) was reductionism, which assumed that science as a whole could be best understood by means of a decomposition of phenomena into smaller, more basic parts (Causey, 1977). Engineering, by extension, was expected to follow the same reductionist trend. The tide of the times in some parts of philosophy was clearly against a systems understanding of technology.

Nonetheless, philosophers did eventually turn to the sociotechnical systems approach by starting not with abstract conceptual models but with empirical observations of how sociotechnical systems were constituted, and then building up. As an exemplar, consider the account by Vermaas, et al. (2011). For them, technical artifacts are physical/social objects that operate within a hybrid system—these are sociotechnical systems. Taking modern aviation as an exemplar, they claim that "[w]hat makes the aviation system a special sort of system is the fact that it includes all kinds of components—the organisations and institutions, the conditions and rules—that are not tangible things... it is a *hybrid* system. It consists of components which, as far as their scientific description goes, belong in very many different 'worlds'" (2011: 68–69). This hybrid nature has an important implication for design of sociotechnical systems: it is no longer the sole province of engineering knowledge, since such knowledge "is no longer adequate when the artefacts attain a form of complexity that leads us to introduce the notion of sociotechnical systems… [that are] less predictable than traditional technical artefacts; sociotechnical systems can display unexpected behaviour even if the end-users set out to use the system in a 'neat' or 'tidy' way" (2011, p. 80). If Vermaas et al. are correct about the proper knowledge required for sociotechnical systems design, another result follows quite directly: explanatory knowledge of human behaviors must not be considered as inferior to knowledge of the physical sciences (engineering included). Or to reduce this conclusion to a slogan: un-tidy design requires un-tidy explanation. When the aviation system loses your luggage, it could be due to a conveyor belt, a software error, a human being who has torn off an ID tag, a lapse in airport management, an inadequate baggage-handling protocol between airlines, a strike due to intransigent management, or any of a number of other factors, human and technological.

One particular ontological complication, mentioned by Vermaas et al., is the "boundary" issue with sociotechnical systems. In the overlap example involving mobile phones and automobiles above, we get a hint of the problem: sometimes it is difficult to know where one system ends and another begins. In placing boundaries, how do we consider integrated phones in an automobile? What about a mobile phone that is used to open car doors and start a car? Have these technologies overlapped in such a way that they cross their original boundaries? Even if this is the case, it would not be so unusual for sociotechnical systems (Stahl, 2021). Consider, for instance, the progression of communication technologies in the U.S. from the middle of the twentieth into the early twenty-first century. In the twentieth century the major provider of home telephone service in the U.S., AT&T, was declared a monopoly in 1984 and broken up into the "Baby-Bells", which soon would provide domestic Internet service (via phone-line modems), cellular service, and would eventually merge with Timer-Warner to provide cable television service and with Verizon to create an Internet service and content provider giant (Hagey et al., 2016).

During that period, the emerging communications giant would, in its various forms, sell to consumers many kinds of technologies and services, be governed by a variety of state and federal regulations, and have customers and user groups that engaged in many different "communications" behaviors. How many systems were there, and where did they begin and end? It is hard to say.

A more recent and fleeting example of a technology with unclear boundaries was the Microsoft chatbot Tay. Lasting less than 1 day, Tay was an AI bot that was trained by Natural Language Processing from users in the U.S., China, and Japan but in its new release relied additionally on Twitter interactions to fill out its conversational abilities. One commentator noted that, once released to the Twitter public, Tay learned language *and* values from Twitter users (Schwartz, 2019). As should have been predicted, this additional training resulted in some controversial racist and sexist utterances from Tay. Soon after Microsoft decommissioned Tay, a Microsoft executive wrote that Twitter users had "exploited a vulnerability" of the Tay language-learning model (Lee, 2016). But that explanation seems to obfuscate the fact that the Twitter community was made an explicit part of the training model—thus it was part of the (short-lived) sociotechnical system that was Tay. If anything, the comment indicates that Microsoft did not understand the boundaries of their own system.

Prediction

For sociotechnical systems ranging from civil aviation to chatbots, it would be helpful if we had predictions of failures. What is the role of technological explanations in providing technological predictions? A straightforward answer is through induction: if a technology fails repeatedly and in the same way, a prediction of its future failure is warranted by the reasonable assumption that the future will resemble the past. But this is by far not the only blueprint for prediction, and it is an implausible one given what we've learned about technological explanations. How else can we produce good predictions?

In an important insight in the ethics of technology, Philip Brey (2012a, 2012b) developed an account of anticipatory ethics that describes how and why we need predictions of failure long before we experience them. He saw the central problem for the ethics of emerging technologies, including information technologies, as a lack of ethical analysis before technologies are entrenched. Once entrenched, technology failures can be costly, dangerous, and hard to root out. But well before their introduction into the marketplace—at the R&D stage—we typically do not know enough about future applications. This is the problem of anticipatory ethics: "How can we do ethical analysis of technologies when we do not have reliable knowledge of their future applications and social consequences? It would seem undesirable that ethicists lose themselves in speculation over future ethical issues that might emerge but that may be based on completely inadequate projections of future developments" (Brey, 2012a, p. 307). Nonetheless, Brey puts faith in forecasting and future studies to provide this knowledge. He grants that this approach "…is more speculative, in that it actually aims to predict or hypothesize devices, uses and social consequences of an emerging technology."

A shortcoming of the anticipatory approach is that, while it focuses on social consequences, it seems not to adequately account for social factors and influences in the ethical analysis of the technology. That is, it locates the social as an effect rather than both cause and effect. To see how, we can consider the three levels at which ethical analysis for Brey can play a role: Technology, Artifact, and Application levels (2012a, p. 310). With respect to the Artifact level, he notes that "[c]ertain complex artifacts, like power plants and railroad systems, may involve human actors as well. In such cases, human actors playing predefined roles are part of the design of the artifact, and the artifact is hence not a completely physical entity but also, in part, a social one." But instead of incorporating the design into the artifact level, it seems more accurate to acknowledge a distinct level—the artifact-as-designed—which could account for the social component of the artifact. The artifact-as-designed might also be identified with an archetype, the last version of a prototype before the artifact is produced. Such a distinction would also help to avoid confusion between failures that are caused by artifact deficiency, for instance, a materials failure in one artifact, as opposed to a design failure that appears in every artifact. Further, the prediction of artifact failures, as in tensile-strength testing of materials, are quite a different matter when compared to a design test of an archetype with a specific user group. This shortcoming notwithstanding, taking the anticipatory approach does lead us to focus on forecasting, which is a part of prediction when failures are not inductive repeats, as mentioned above. Moreover, the anticipatory approach could help us to avoid both underestimating and overestimation of ethical problems with future applications of critical technologies, e.g., artificial intelligence (Powers & Ganascia, 2020).

To make good on the anticipatory approach, we will need a predictive model for sociotechnical systems behavior. In most cases, this will present a daunting task. In thinking about the examples that we have considered so far, there have been instances of direct, indirect, looping, and non-linear causation; technological efficiencies that are undone by psycho-social factors; technological problems that can augment one another and are neither clearly social nor clearly technical; sociotechnical systems that have vague boundaries; and corporate and political motivations that obscure why a technology has failed. We can take as our starting point the mathematical models of the physical sciences, though it is apparent that they will only get us part of the way towards modeling a sociotechnical system. A mathematical model of a physical system predicts future behavior of parts of a system by means of equations. Such equations require that the data for building the model, as well as using the model to make predictions, are quantitative or can be approximately quantifiable where they are originally qualitative. We can consider a carbon cycle-climate model as an example (Strassmann & Joos, 2018). With a series of equations derived from historical data, the model can predict with some accuracy the behaviors of the climate-carbon system, including surface air temperature change, given input data such as atmospheric carbon dioxide levels. Computation over these equations, as Strassman et al. point out, can give us predictions with expressed degrees of uncertainty. The more data we have, the better the models become, since each data point is in principle an avenue towards correcting an equation.

For predictive models of sociotechnical systems, some of the relations that we would want to express are indeed quantifiable—we can make equations out of them—though many others will be qualitative and stubbornly so. For instance, the regulatory relationship between product testing standards, enforcement of the standards, and industry best practices is a complicated tale. Once we observe a system in sufficient detail, we may be able to express (qualitatively) what that relationship is. It is extremely doubtful, however, that all relationships between all parts of a sociotechnical system can be described, and where they can, it is even more doubtful that they can be quantified. Given the argument so far, that explanations are important but complicated, and that predictions are quite desirable but cannot simply be read off of past explanations of failure, what can we do?

The Computational Turn: Promise and Peril

An intermediate conclusion can be drawn from these attempts by scholars in organizational studies, engineering, and philosophy: Satisfactory explanation and prediction of technological behaviors have so far eluded human understanding. Perhaps then we might make more progress by turning to a technology itself to better understand sociotechnical systems. Here then we consider a "computational turn" to the sociotechnical approach by introducing cases of knowledge generation by means of machine learning models and Knowledge Graphs (Chaudhri et al., 2022). The objective of this turn is to sketch a possible application of computational technologies with which to more accurately express relationships between sociotechnical entities and relations within a system.

With the recent interest in machine learning (ML) models, it might be enticing to think that ML could help us understand sociotechnical systems. While there are several kinds of ML and some work well on unstructured data—what we could now make available regarding some sociotechnical systems, such as relevant text and images—it is unclear how useful the resulting models would be. No doubt, usefulness would depend in part on the size and quality of the data. Some researchers have applied traditional ML methods to understand failures in information systems, but with inconclusive results (Strang & Vajjhala, 2022). Others are optimistic about Deep Learning methods, which have been implemented with some success on unstructured data. And foundational Large Language Models such as GPT-4 could eventually be integrated with application models for sociotechnical systems. It is difficult to say exactly how far this technology will progress, even in the near future. One factor in future implementations seems to be the profit motive. While sociotechnical systems theory has so far been the province of academics, a reliably predictive computational model of a sociotechnical system could have tremendous benefits for companies hoping to introduce new technologies and minimize risk. This has already been done on a smaller scale, with so-called in silico models for drug and medical device testing. The models that have been developed are being used by some companies to present computed data as evidence of safety and/or efficacy in place of clinical trials. Lack of validation of the models has led the U.S. Food and Drug Administration to implement an oversight program on the "credibility of computational models" for such uses (USFDA, 2023). But again, the lesson here is that, if there is a business case for applying ML to sociotechnical systems, some corporations are likely to attempt it.

A less ambitious computational approach would be to construct a kind of data model called a Knowledge Graph (KG) (Ehrlinger & Wöß, 2016), which integrates information about entities into an ontology. KGs use structured data (e.g., Wikidata) and express searchable relationships between entities. In this sense, developers of KGs often say that the model can "reason" about the information in the database in that it can return true inferences when a person types in a query. While the technology is barely a decade old, Google has used a KG since 2012 as part of Google Search, to provide a new "semantic search" feature that relays information via the infobox next to search results. Singhal (2012) announced the Google KG with some hype, promising a new semantic search that would deliver "things, not strings." By 2020, the Google KG had grown to express over 500 billion facts about 5 billion entities (Sullivan, 2020). Initially, KGs used binary relations between two entities—a triple [Subject-Predicate-Object] structure—to describe such facts. Recent versions of KGs are also known as Hyper Graphs, which can express n-ary relations between entities (Chen et al., 2022). One reason that KGs are attractive for our prediction goal is that "[t]hey are also beginning to play a central role in representing information extracted by AI systems, and for improving the predictions of AI systems by giving them knowledge expressed in KGs as input" (Chaudhri et al., 2022).

Criticisms of a computational turn for sociotechnical systems are likely to be several, and maybe even severe. In general, there is amongst sociotechnical systems theorists a healthy distrust of AI, as many AIs are thought (sometimes mistakenly) to be "black boxes," the behavior of which can be neither explained nor understood. There would be a certain irony if this turned out to be the case with a computational approach to understanding sociotechnical systems. Oddly, we would have used AI systems to better understand sociotechnical systems, but then we (humans) could not understand what the AI had learned about such systems. A deeper skeptical worry comes with the realization that an AI understanding of a sociotechnical system would depend directly on the data on which it was trained. If this data misrepresents a particular sociotechnical system, the AI would not represent what AI researchers call "ground truth" but merely the patterns and relationships in the data. As with all AI systems that learn from data, what they learn cannot connect with an underlying reality if the data do not themselves correspond to that reality in some way.

Expanding on the skeptical worries, we should consider whether the computational turn for sociotechnical systems would follow in the footsteps of a broader movement of the late twentieth century known as computational data science (CDS).

Several writers have speculated that, if the trend in big-data "discovery science" continues, the results can only be that we produce a science that computers but no humans can understand. CDS is the application of data science techniques, often involving the building of machine learning models, to understand our natural and social worlds. The main worry is that CDS leaves humans behind as creators of science—and thus that we will become bystanders to scientific knowledge. The root of the worry is that there is a disconnect between what was previously considered to be the gold standard of science—causal knowledge of the natural world—and the "mere" statistical knowledge that most of CDS generates. One such writer, David Weinberger (2011), worries that CDS is on the verge of making scientific understanding impossible for humans because of the difference between the "internal" details of such scientific knowledge, accessible only to the computer, in contrast to what the computer can make understandable to humans through whatever interface they select.

A somewhat dystopian worry about the computational sociotechnical systems approach is that an AI-powered predictive model would lend itself to social engineering. Prediction is typically seen as normatively neutral or beneficial, as in our desired predictions of technology failure that motivate this inquiry. But if the central thesis of sociotechnical systems theory is correct—that human behavior and social institutions interact with technologies—it cannot be ruled out that computational predictions could be used to manipulate human behaviors. Indeed, Facebook has already produced research on the power of emotional contagion through manipulation of their Newsfeed (Kramer et al., 2014), and there is more recent evidence that Facebook executives have doubled down on algorithms that produce profitable "engagement" on the platform at the expense of psychologically healthy online interactions (Hagey & Horwitz, 2021).

The Accountability Challenge for Computational Systems

Whether a predictive AI that was designed originally to better forecast behavior of a sociotechnical system could be used to manipulate people is a somewhat science-fiction scenario—much like the Terminator was designed to protect human society but ended up threatening it. However, a quite real and present threat comes with AIs that control aspects of sociotechnical systems—think of Google's Search or the navigation algorithm of an autonomous vehicle. These AIs may be difficult to hold to account in the event of both "satisfaction" and "expectation" failures, but at the same time, their behaviors could have serious consequences for human life and wellbeing. To return to the example from the Introduction, consider that in recent years AI systems have been deployed that classify and recommend within our penal system in ways that are unfair or biased for certain populations, and there are an increasing number of examples in which such underperformance exhibits the equivalent of technological racial or gender discrimination on both empirical (quantifiable) and conceptual grounds (Scantamburlo, 2021). In cases such as these, we need conceptual clarity on who or what to hold to account. In "Algorithmic Accountability in the Making" (2021) Deborah Johnson considers whether an algorithm can be the subject of such accountability. Johnson's argument is an example of a sociotechnical systems approach in that she considers algorithms and accountability in the broader context in which they both operate, namely, in current social contexts in which persons (and sometimes corporations) are held to account mostly through the legal system, but artifacts and technological systems are not.

Johnson resists ascribing accountability to algorithms when the focus is on the algorithm itself—for instance, on the mere transparency of its operations. In other words, whether the operations are transparent or not will not settle the question of whether it can be held accountable for a failure. Instead of looking merely at the algorithm, she turns our attention to the inherently social aspect of "holding accountable." Following Bovens (2007), she describes the social practice of accountability as involving actors who share a belief that an account of some behavior is called for, and in addition, that this accounting takes place in a forum where there are procedural norms for such events. On her view, accountability for algorithms is dubious precisely because algorithms "are not capable of having beliefs and obligations, not capable of giving explanations of their behavior, responding to interrogation, and facing consequences" (Johnson, 2021, p. 124). Algorithms alone do not believe, have obligations, or offer up explanations or responses. Thus, for her, their meaningful inclusion in the practice of accountability is unlikely.

Further considerations about the practice of accountability are in order here, and may suggest that in the not-so-distant future, algorithms may indeed be held to account. First, let us consider one type of "holding to account" that takes place in a court of law. In a typical civil proceeding, defendants are summoned to court, read the complaints against them, apprised of their rights, and must be present for the hearing during which arguments for and against several kinds of legal accountability are aired. After the proceedings, which follow prescribed rules of evidence and testimony, cross-examination, and appeal, a judge or jury may deliver various remedies requiring the payment of damages. But in addition to damages, so-called equitable remedies, including injunctions to cease and desist, may be handed down as a judgement. Natural persons are often the beneficiaries of damages, and they can be required to pay damages. Both payment of damages and being subject to injunctions are forms of accountability. Importantly, these verdicts do *not* require a determination that the person who is held to account intended any of the damages. All that is required is that they behaved in a certain way—or perhaps failed to behave as they should have. This is a significant departure from criminal law, where conviction requires showing intent of those convicted of crimes—it requires a projection about the mind of the defendant. Thus, the law acknowledges and enforces accountability for natural persons in civil cases, but does not here attempt to "enter" the mind of the defendant.

Second, in addition to this treatment by the civil law of natural persons, corporations and universities in the U.S. have been deemed artificial persons in a slowly-evolving body of case law (Horwitz, 1986). Like natural persons, artificial persons can be subject to financial or equitable remedies and injunctions, but they do not enjoy all of the rights of natural persons, e.g., they do not have the 5th Amendment right not to testify against oneself. Unlike natural persons, but *very* much like algorithms, corporations and universities are not capable of having thoughts or beliefs. So, in consideration of the complaints against a corporation, a judge or jury does not consider corporate "intent"—and it is rather difficult to understand what this would even mean. Both natural and artificial persons can be represented in a court of law by counsel, or a guardian, when they are held accountable. Fenwick and Wrbka (2022) also deploy the analogy between algorithms and artificial persons to suggest that, at least as far as legal standing goes, an algorithm could be subject to the type of "holding to account" which judges and juries deliver.

Much like the formative question of environmental law in the classic "Should trees have standing?" (Stone, 1972), the case for algorithmic accountability need not turn on consciousness, cognition, or self-expression. If this is correct, it would seem that Johnson's position is more accommodating than perhaps she intended—that it could make room for algorithmic accountability—given that there is indeed a social (i.e., legal) practice which at least foreshadows future accountability, even though clearly this is not the intended purpose of her analysis. No doubt, she is right that standard examples of accountability in practice involve linguistic and cognitive performances on the part of those being held to account. It is a further question whether all accountability practices must include linguistic and cognitive performances.

A final consideration in favor of algorithmic accountability recalls the distinction between justifying and non-justifying explanations mentioned earlier. Here we must consider that distinction as it applies to two quite different computational systems: standard, presently available software packages like tax-preparation software, and cutting-edge, learning AI systems that pursue goals through means that are often not specified. The first kind of computational system can be considered a surrogate agent (Johnson & Powers, 2008) which is capable of carrying out the interests of the client, but not having (first-person) interests itself. In our consideration of surrogate agents we refrained from claiming that such surrogate agents were responsible or could be held liable; at most, it seems these examples reserve accountability for the maker of the software, or perhaps in some rare cases for the user. If an incorrect formula was used in the software to calculate a client's tax bill, it would be wrongheaded to hold the software accountable. Upon discovering that an incorrect formula was used, the (human) client may well offer that formula as an explanation to the tax authority, but the explanation—accurate as it may be—would not justify the client being allowed to pay less than what was owed. At the end of the day, the client must justify—to the taxing authority—why software calculations were not validated (US Tax Court, 2017). For computer surrogate agents, justifying still resides with humans.

The second kind of computational system may be presently available or soon will be—systems that interpret human directives through AI tools such as Natural Language Processing and Large Language Models and carry them out through what *the system* has computed to be the most efficient means, according to the model on which they operate. Early examples in the autonomous vehicle industry include vehicles that would park and navigate to a particular destination without further human input. Surely, these systems have failed in some dramatic cases and explanations of those failures alone are helpful in making them safer. In this respect, these AIs are like older software systems. What sets them apart, however, is their ability to choose means to effect certain ends, where the choice of those means would typically call for not only an explanation but a justification *by a human agent.* Autonomous vehicles are not in the least the most worrisome cases of these seemingly accountable technologies. Already in wars in the early twenty-first century there is evidence that AI battlefield systems are deciding who to kill and how to kill (Pratt, 2024). If we do end up asking for justifications from our AI systems, it will be a small (conceptual) step to go further and hold them accountable.

A thorough sociotechnical explanation of AIs as implemented—one that looked at algorithm operation and creation and degree of autonomy in selecting means to an end—could contribute to a decision by regulatory or legal authorities to hold someone or *something* to account. Depending on where the inquiry stops, several entities could be assigned some degree of accountability: a designer or design team, a CEO of the corporation that makes the technology, a government official with oversight authority, or perhaps an algorithm. In an example of injunctive relief, the algorithm could be banned from future implementations. When we successfully explain behaviors in an AI sociotechnical system, we learn what human actors, what organizations, what laws and regulations, and yes—what algorithms—caused the failure in question. Absent such an explanation, there can be no definitive case either for or against accountable algorithms.

Conclusion

Sociotechnical explanations can contribute to individual failure investigations and even normative critiques of an entire system, in the event that clear descriptions of causes and conditions of failure can be specified. But in many failed systems, what has happened is complicated and our understanding of them is confused. At present, sociotechnical systems approaches in organizational studies, engineering, and philosophy do not meet standards of clarity that are required of explanation. Further, sociotechnical system approaches so far have not been used to predict instances of failure, though it would be highly desirable if we could develop such predictions. Complications with explanation and the dim prospects for prediction may lead us to seek assistance from new computational technologies, but reliance on such technologies is not without risks. One problem is that the explanations and predictions that we gain by using computers for such epistemic tasks could (ironically) elude our meager understanding. On the other hand, technological development seems not to be hindered by our limited human understanding of how best to exploit our own technologies. A special case of this problem may well arise with the problem of algorithmic accountability. While our practice, heretofore, with "holding to account" has been an entirely human affair, and thus our concept of accountability has taken the shape of the mold provided by humans, it may well be that the mold is changing radically, since new kinds of agents are making decisions formerly reserved for us alone.

References

Bovens, M. (2007). Analysing and assessing accountability: A conceptual framework. *European Law Journal, 13*(4), 447–468.

Brey, P. (2012a). Anticipating ethical issues in emerging IT. *Ethics and Information Technology, 14*, 305–317. https://doi.org/10.1007/s10676-012-9293-y

Brey, P. (2012b). Anticipatory ethics for emerging technologies. *Nanoethics, 6*(1), 1–13. https://doi.org/10.1007/s11569-012-0141-7

Causey, R. L. (1977). *Unity of science*. Reidel.

Chaudhri, V. K., Baru, C., Chittar, N., Dong, X. L., Genesereth, M., Hendler, J., Kalyanpur, A., Lenat, D., Sequeda, J., Vrandečić, D., & Wang, K. (2022). Knowledge graphs: Introduction, history, and perspectives. *AI Magazine, 43*, 17–29. https://doi.org/10.1002/aaai.12033

Chen, Z., Wang, X., Wang, C., & Li, J. (2022). Explainable link prediction in knowledge hypergraphs. In *Proceedings of the 31st ACM international conference on Information & Knowledge Management* (CIKM '22). Association for Computing Machinery, New York, NY, USA (pp. 262–271). https://doi.org/10.1145/3511808.3557316

Davis, M. C., Challenger, R., Jayewardene, D. N. W., & Clegg, C. W. (2014). Advancing sociotechnical systems thinking: A call for bravery. *Applied Ergonomics, 45*(2A), 171–180. https://doi.org/10.1016/j.apergo.2013.02.009

Dressel, J., & Farid, H. (2018). The accuracy, fairness, and limits of predicting recidivism. *Science Advances, 4*, 1. https://doi.org/10.1126/sciadv.aao5580

Ehrlinger, L., & Wöß, W. (2016). Towards a definition of knowledge graphs. *International conference on semantic systems* (Vol. 1695).

Eisenstein, E. (1980). *The printing press as an agent of change*. Cambridge University Press. https://doi.org/10.1017/CBO9781107049963

Fenwick, M., & Wrbka, S. (2022). AI and legal personhood. In L. DiMatteo, C. Poncibò, & M. Cannarsa (Eds.), *The Cambridge handbook of artificial intelligence: Global perspectives on law and ethics* (Cambridge Law Handbooks: pp. 288–303). Cambridge University Press. doi:https://doi.org/10.1017/9781009072168.028

Gitlin, J. (2023). Hackers discover that vulnerabilities are rife in the auto industry. *Ars Technica* January 11, 2023. Retrieved from https://arstechnica.com/cars/2023/01/hackers-discover-that-vulnerabilities-are-rife-in-the-auto-industry/

Gunderson, R. (2017). The problem of technology as valuation errors: The paradox of the means in Simmel and Scheler. *Technology in Society, 48*, 64–69. https://doi.org/10.1016/j.techsoc.2016.12.005

Hagey, K., & Horwitz, J. (2021). Facebook tried to make its platform a healthier place. It got angrier instead. *The Wall Street Journal* Sept. 15, 2021.

Hagey, K., Sharma, K., Cimilluca, D., & Gryta, T., (2016). AT&T is in advanced talks to acquire Time Warner, *The Wall Street Journal* Oct. 21, 2016.

Hempel, C. G. (1965). *Aspects of scientific explanation and other essays in the philosophy of science*. Free Press.

Hempel, C. G., & Oppenheim, P. (1948). Studies in the logic of explanation. *Philosophy of Science, 15*(2), 135–175. https://doi.org/10.1086/286983

Horwitz, M. J. (1986). Santa Clara revisited: The development of corporate theory. *West Virginia Law Review, 88*(2), 173–224. Retrieved from https://researchrepository.wvu.edu/wvlr/vol88/iss2/5

Johnson, D. G. (2021). Algorithmic accountability in the making. *Social Philosophy and Policy, 38*, 111–127. https://doi.org/10.1017/S0265052522000073

Johnson, D. G., & Powers, T. M. (2005). Computer systems and responsibility: A normative look at technological complexity. *Ethics and Information Technology, 7*(2), Kluwer Publishing.

Johnson, D. G., & Powers, T. M. (2008). Computers as surrogate agents. In M. J. van den Hoven & J. Weckert (Eds.), *Information technology and moral philosophy*. Cambridge University Press.

Johnson, D. G., & Verdicchio, M. (2017). Reframing AI discourse. *Minds and Machines, 27*(4), 575–590.

Kollnig, K., Shuba, A., Van Kleek, M., Binns, R., & Shadbolt, N. (2022). Goodbye tracking? Impact of iOS App tracking transparency and privacy labels. In *Proceedings of the 2022 ACM conference on fairness, accountability, and transparency (FAccT '22)*. Association for Computing Machinery, New York (pp. 508–520). https://doi.org/10.1145/3531146.3533116

Kramer, A. D., Guillory, J. E., & Hancock, J. T. (2014). Experimental evidence of massive-scale emotionale contagion through social networks. *Proceedings of the National Academy of Sciences, 111*(24), 8788–8790. https://doi.org/10.1073/pnas.1320040111

Law Library Of Congress, U. S. G. L. R. D. (2008). *Shutter sound of cell phone cameras: Japan & South Korea*. The Law Library of Congress, Global Legal Research Directorate. [Pdf] Retrieved from the Library of Congress. https://www.loc.gov/item/2019669583/

Lee, P. (2016, March 25). Learning from Tay's introduction. Official Microsoft Blog. *Microsoft*. Retrieved from https://blogs.microsoft.com/blog/2016/03/25/learning-tays-introduction/

Leveson, N. G. (2005). A systems-theoretic approach to safety in software-intensive systems. *IEEE Transactions on Dependable and Secure Computing, 1*(1), 66–86. https://doi.org/10.1109/TDSC.2004.1

Marcuse, H. (1964). *One-dimensional man*. Beacon Press.

McCann, W., Pedneault, A., Stohr, M. K., & Hemmens, C. (2018). Upskirting: A statutory analysis of legislative responses to video voyeurism 10 years down the road. *Criminal Justice Review, 43*(4), 399–418. https://doi.org/10.1177/0734016817741342

Miller, C., & Valasek, C. (2015). Remote exploitation of an unaltered passenger vehicle. *Black Hat USA*. Retrieved from https://ioactive.com/pdfs/IOActive_Remote_Car_Hacking.pdf

Mumford, E. (2006). The story of socio-technical design: Reflections on its successes, failures and potential. *Information Systems Journal*, 16(4), 317–342. Retrieved from https://doi.org/10.1111/j.1365-2575.2006.00221.x

Ortega, C. A. C., Mariscal, M. A., Boulagouas, W., Herrera, S., Espinosa, J. M., & García-Herrero, S. (2021). Effects of mobile phone use on driving performance: An experimental study of workload and traffic violations. *International Journal of Environmental Research and Public Health., 18*(13), 7101. https://doi.org/10.3390/ijerph18137101

Porter, A. (2023). Girls aged three among victims of upskirting. *Sky News*, August 31, 2023. Retrieved from https://news.sky.com/story/girls-aged-three-among-victims-of-upskirting-as-teenager-describes-anger-at-schools-response-12949592

Powers, T. M., & Ganascia, J.-G. (2020). The ethics of the ethics of AI. In M. D. Dubber, F. Pasquale, & S. Das (Eds.), *The Oxford Handbook of Ethics of AI*. Oxford Academic. https://doi.org/10.1093/oxfordhb/9780190067397.013.2

Pratt, S. F. (2024). When AI decides who lives and dies. *Foreign Policy*, May 2, 2024. Retrieved from https://foreignpolicy.com/2024/05/02/israel-military-artificial-intelligence-targeting-hamas-gaza-deaths-lavender/

Rahmillah, F. I., Tariq, A., King, M., & Oviedo-Trespalacios, O. (2023). Is distraction on the road associated with maladaptive mobile phone use? A systematic review. *Accident Analysis & Prevention, 181*(106900). https://doi.org/10.1016/j.aap.2022.106900

Richardson, A. V. (2020). *Bearing Witness While Black: African Americans, smartphones, and the new protest #journalism* (New York, 2020; online edition, Oxford Academic, 18 June 2020). https://doi.org/10.1093/oso/9780190935528.001.0001

Salmon, W. C. (1971). Statistical explanation. In Salmon (Ed.), *Statistical explanation and statistical relevance* (pp. 29–87). University of Pittsburgh Press.

Scantamburlo, T. (2021). Non-empirical problems in fair machine learning. *Ethics and Information Technology, 23*, 703–712. https://doi.org/10.1007/s10676-021-09608-9
Schwartz, O. (2019). In 2016, Microsoft's racist Chatbot revealed the dangers of online conversation. *IEEE Spectrum*, November 25, 2019. Retrieved from https://spectrum.ieee.org/in-2016-microsofts-racist-chatbot-revealed-the-dangers-of-online-conversation
Singhal, Amit (2012). *Introducing the knowledge graph: Things, not strings Google Blog* May 16, 2012. Retrieved from https://blog.google/products/search/introducing-knowledge-graph-things-not/
Srivastava, L. (2005). Mobile phones and the evolution of social behaviour. *Behaviour & Information Technology, 24*(2), 111–129. https://doi.org/10.1080/01449290512331321910
Stahl, B. (2021). *Artificial intelligence for a better future: An ecosystem perspective on the ethics of AI and emerging digital technologies*. Springer Nature.
Stone, C. (1972). Should trees have standing? Towards legal rights for natural objects. *Southern California Law Review, 450.*
Strang, K. D., & Vajjhala, N. R. (2022). Applying machine learning methods to understand unstructured information system big data. In *2022 17th Iberian Conference on Information Systems and Technologies (CISTI)*, Madrid, Spain, 2022, (pp. 1–6). https://doi.org/10.23919/CISTI54924.2022.9820116
Strassmann, K. M., & Joos, F. (2018). The Bern Simple Climate Model (BernSCM) v1.0: an extensible and fully documented open-source re-implementation of the Bern reduced-form model for global carbon cycle–climate simulations. *Geoscientific Model Development, 11*, 1887–1908. https://doi.org/10.5194/gmd-11-1887-2018
Sullivan, D. (2020. A reintroduction to our knowledge graph and knowledge panels. *Google blog*. May 20, 2020. Retrieved from https://blog.google/products/search/about-knowledge-graph-and-knowledge-panels/
Trist, E. L., & Bamforth, K. W. (1951). Some social and psychological consequences of the longwall method of coal-getting: An examination of the psychological situation and defenses of a work group in relation to the social structure and technological content of the work system. *Human Relations, 4*(1), 3–38. https://doi.org/10.1177/001872675100400101
U.S. Food and Drug Administration. (2023). *Credibility of computational models program: Research on computational models and simulation associated with medical devices*, 11/16/2023. Retrieved from https://www.fda.gov/medical-devices/medical-device-regulatory-science-research-programs-conducted-osel/credibility-computational-models-program-research-computational-models-and-simulation-associated
U.S. Tax Court. (2017). Bulakites v. Commissioner of Internal Revenue, Tax Court Memo. 2017-79 (U.S.T.C. May. 11, 2017).
Vermaas, P., Kroes, P., Poel, I. v. d., Franssen, M., & Houkes, W. (2011). *A philosophy of technology: From technical artefacts to sociotechnical systems*. Springer International Publishing. ISBN: 9783031799716.
Weinberger, D. (2011). *Too big to know*. Basic Books.

The manufacturer's authorised representative in the EU is Springer Nature Customer Service Centre GmbH, Europaplatz 3, 69115 Heidelberg, Germany. If you have any concerns regarding our products, please contact ProductSafety@springernature.com

Printed and bound by CPI Group (UK) Ltd, Croydon, CR0 4YY
30/09/2026
02225608-0018